BUS REVIEW 15

Review of 1999

Published 2000

Above: Looking to the future, FirstGroup worked closely with Wright of Ballymena in the development of an exciting new range of bodies. This is the Eclipse Fusion on Volvo B7LA chassis which was the star of Coach & Bus 99.

Front cover: Seven single door Plaxton President-bodied Dennis Tridents were added to the Lothian fleet for its relaunched Airlink service. Dual doored Tridents with both Alexander and Plaxton bodies were bought in 1999. *John Burnett*

Back cover, upper: A-Bus of Bristol was the first British operator of a low-floor double-decker, in early 1998. In 1999 it added this Alexander-bodied DAF to its fleet.

Back cover, lower: Unusual vehicles for a small fleet - even if it is part of the EYMS Group - were two Mercedes-Benz O405Ns which entered service with Finglands of Manchester in August.

Contents

The last old-style step-entrance double-deckers for London were 20 Volvo Olympians for CentreWest. They were also the last Northern Counties bodies, as production at Wigan switched to the Plaxton President (see page 41).

First published 2000
ISBN 0–946265–32–1

Typeset in Times and Helvetica
Electronic page makeup by Jeremy Scott
Printed by Pillans & Wilson, Edinburgh

Published by
Bus Enthusiast Publishing Company
5 Hallcroft Close,
Ratho, Newbridge
Midlothian
EH28 8SD
www.busenthusiast.com
Bus Enthusiast is an imprint of
Arthur Southern Ltd.

Introduction

THERE WAS a time when the London bus scene was relatively uninteresting, with London Transport running a highly-standardised fleet and buying buses by the hundreds. Go back 20 years and these buses were Titans and Metrobuses; before that there were Fleetlines and Swifts; before that Routemasters ... and so it goes on.

But in 1999 London was the place to be, as a new generation of low-floor double-deckers took to the capital's streets in ever increasing numbers. Most were Dennises. Among the operators, too, there was change - London Traveller and Mitcham Belle were new names on tendered services - and while Limebourne bounced back from its troubles when Q-Drive collapsed in 1998, the year ended with Harris Bus in administrative receivership.

Elsewhere, articulated buses were in the news in some of the country's biggest conurbations. They arrived in Birmingham, Glasgow, Leeds and Manchester - and departed in Sheffield.

There was significant overseas expansion by British operators and manufacturers. FirstGroup, National Express and Stagecoach all made significant acquisitions in the United States where between them they now control an amazing 25,000 vehicles ranging from rugged yellow school buses to high-specification touring coaches.

Among the manufacturers, Henlys has the strongest presence through its acquisition of Blue Bird, which it adds to its joint ventures with Volvo in North America. Mayflower, which has a significant minority stake in Metrotrans (which filed for bankruptcy at the end of the year), set up a joint operation with Thomas Built Buses which will see Alexander-bodied Dennis Darts being manufactured in North Carolina. And Wright of Ballymena is working with Chance Coach to supply buses to the North American market.

Back home, Volvo launched its new B7L, a model intended to combat the supremacy which Dennis now has in the supply of double-deck buses. From Dennis came a new R-series coach chassis, although production doesn't get underway until well into 2000. On the other hand Plaxton's new coach range was in build at the end of the year and significant orders were announced - over 100 in all - for delivery in 2000.

Wright stole the show at Coach & Bus 99 with its new Solar and Eclipse models. The names were topical as Britain experienced a solar eclipse in the summer, which briefly left parts of the West Country in daytime darkness.

The corporate colours of Arriva spread rapidly during 1999, while those of FirstGroup depended on the appearance of new buses - but there were plenty of those.

Indeed new bus deliveries in 1999 totalled 1,863 according to the Society of Motor Manufacturers and Traders, up 68 per cent on the 1,112 delivered in 1998. Coach sales were down slightly to 1,049, from 1,127 in the previous year.

Dennis was bus market leader accounting for half of all new bus sales and selling four times as many buses as its nearest rival, Volvo. Volvo, on the other hand, retained its leadership of the coach market, with a share of just over 40 per cent. Scania was the number two in coach sales.

To keep up with unfolding events, *Buses* magazine provides informed news and views every month, while the PSV Circle produces monthly news-sheets which record in remarkable detail who is buying what from whom. Current fleet lists of major operators - and many minor ones - are to be found in the regularly-updated Bus Handbook series produced by British Bus Publishing.

There is also a vast amount of information on the internet - some bits of it more reliable than others - and where sections of the review deal with a specific operator or manufacturer we have included this year for the first time their website addresses. Bus Enthusiast publishing has also embraced electronic technology and can be visited at *www.busenthusiast.com.*

Stewart J Brown
Reedley Hallows, 2000

An impressive new bus for Norfolk Green was this Mercedes-Benz O405N. ***Geoff Mills***

New in the North West

The first-ever Dennises for Preston Bus or its predecessors were seven Tridents in December. They had East Lancs bodies and were used on a route in which the operator and local authority had formed a Quality Partnership. The Tridents are fitted with transponders to give them priority at traffic lights - turning red lights to green, or holding green phases to let the bus through.

On certain high-profile contracts Merseytravel - the Merseyside PTE - specifies the use of yellow Smart Eco livery. It is seen on a Mini Pointer Dart operated by Supertravel of Speke. The bus incorporates a vehicle location system linked to the provision of real-time service information at stops along its route.

Bluebird of Manchester regularly invests in new buses. Its 1999 intake included a Wright-bodied Dennis Dart.

UK North, which at the start of the year was operating a fleet made up primarily of elderly Leyland Atlanteans, took delivery of four new Dennis Tridents in August. They had Alexander ALX400 bodies. *David Barrow*

Mayne of Manchester joined the growing number of low-floor double-deck operators with five Dennis Tridents with East Lancs bodies. They mark a switch from Scania, which has supplied most of Mayne's new 'deckers in recent years.

A complete break with tradition at council-owned Blackburn Transport saw the delivery of five Volvo B10BLEs with Wright bodies. They carried route branding for the trunk Accrington to Darwen service, which later in the year came under attack from new operator Northern Blue.

Stagecoach expands abroad

STAGECOACH CONTINUED TO expand in 1999, although with few opportunities in the UK the group was looking overseas.

First, in the Far East, it bought Hong Kong Citybus in March. HKCB - one–time owners of Capital Citybus in London - operates 1,187 buses, the majority of which are British-built double-deckers. To this was added in May a 45 per cent stake in Chinese operator Hong Kong Kwoon Chung (Chongquing) which operates 576 buses in Chongquing.

Among Stagecoach's first moves at HKCB was the repatriation to the UK of 21 Dennis Darts with air-conditioned Plaxton Pointer bodies. By the end of the year a few had been allocated to Stagecoach Devon.

On the other side of the globe Stagecoach made its biggest single acquisition ever, with the take-over in June of Coach USA which operates a fleet of some 6,500 buses and coaches across the US. At the same time, and a bit closer to home, it announced that it was taking a 35 per cent stake in the Italian Sogin Group, running 1,800 buses under the Sita name. As the year progressed Coach USA expanded by the acquisition of a number of smaller businesses, but in contrast the Sogin deal foundered.

However against all this expansion, the group announced in October that it was selling Swebus, with completion of the sale to a Nordic company expected at the start of 2000. Swebus had been in Stagecoach ownership since October 1996. The sale was designed to allow the group to focus more of its attention on North America. It left Portugal as the only mainland Eurpean country in which Stagecoach has a presence.

By contrast to its global changes, moves in the UK were tame with only one take-over of note. In February Stagecoach Manchester purchased the 21-vehicle Glossopdale business. Glossopdale had started operations in 1992.

For its London operations Stagecoach bought new Dennis Darts and Tridents. At the end of the year Stagecoach East London was taking delivery of 66 Darts with Plaxton Pointer 2 bodies.

New vehicles for Stagecoach companies in the UK were predominantly Alexander-bodied, and included ALX200s, ALX300s and ALX400s. The ALX300s were all on MAN chassis and appeared in Fife, Manchester and Newcastle-upon-Tyne. The ALX400s introduced Dennis Tridents to the group - initially in London but later in the year in Manchester, Newcastle and Grimsby. Winning Stagecoach double-deck business was a remarkable coup for Dennis. Previously all of the group's double-deck requirement had been met by Volvo and, before that, Leyland. The London Tridents had two-door bodies while those for provincial fleets were single-door buses.

Plaxton supplied dual-door Pointer 2s on Dennis Dart chassis for operation in London, along with a handful of single-door Pointers for operation elsewhere. These included specially-liveried buses for a bus-rail link connecting with Virgin train services at Macclesfield.

Interesting additions to the coach fleet were six articulated Volvo B10Ms with Jonckheere Modulo bodies, and 27 three-axle MAN 24.350s with 68-seat double-deck Jonckheere Monaco bodies. The MANs were used to replace the entire Oxford Tube fleet, running between Oxford and London and reportedly cost £5.5million, equal to just over £200,000 per vehicle. They entered service during the summer. The 72-seat artics were divided equally between Stagecoach Transit (for the Hull to Sheffield service) and Stagecoach Western Buses.

Looking to the future, Stagecoach was named at the end of the year as the preferred operator for the proposed £50 million guided busway in Northampton, an ambitious scheme being promoted by Rapid Transit International.

With around 7,500 buses in the UK, Stagecoach comes in third place after FirstGroup and Arriva, but add its massive operations worldwide and it is by far the biggest of the British-based bus groups.

website:
www.stagecoachholdings.com

STAGECOACH GROUP SUBSIDIARIES

Main companies and trading names:

Bayline
Stagecoach Devon
Bluebird Buses
  Inverness Traction
  Stagecoach
Burnley & Pendle Transport
Busways Travel Services
  Magic Bus
  Stagecoach Busways
Cambus group
  Cambus
  Viscount Bus & Coach
Circle Line
Cleveland Transit
  Stagecoach Darlington
  Stagecoach Hartlepool
  Stagecoach Kingston-upon-Hull
  Stagecoach Transit
Coach USA
Devon General
East London Bus & Coach Co
East Midland Motor Services
  Chesterfield Transport
Eastbourne Bus Company
  (New Zealand)
Fife Scottish Omnibuses
Greater Manchester Buses South
  Magic Bus
  Stagecoach Manchester
Grimsby-Cleethorpes Transport Co
Hong Kong Citybus
Hong Kong Kwoon Chung
  (Chongquing) (45 per cent)
Magicbus Scotland
  (Holding company)
National Transport Tokens (99.9 %)
Parfitt's Motor Services
PSV Claims Bureau
Rhondda Buses
Ribble Motor Services
Scottish Highway Express
Sistema Metrobus de Bogota (25 %)
South East London & Kent Bus Co
  Stagecoach Selkent
Stagecoach Australia
Stagecoach Devon
Stagecoach East Kent
  South Coast Buses
Stagecoach Glasgow
Stagecoach Graphics
Stagecoach International Services
Stagecoach North West
  Cumberland Motor Services
  Coachlines
  Lakeland Experience
Stagecoach Lancaster
Stagecoach Portugal
Stagecoach Scotland
Stagecoach South
  Hampshire Bus
  Stagecoach Hants & Surrey
  Sussex Coastline
Stagecoach Wellington
Stagecoach West
  Aberdare Bus Co
  Cheltenham & Gloucester
    Omnibus Co
  Gloucester Citybus
  Metro
  Stroud Valleys
  Cheltenham District Traction
  Midland Red (South)
    G&G Travel
  Red & White Services
  Swindon & District
Thames Transit
  The Oxford Tube
  Stagecoach Oxford
United Counties Omnibus Co
  Coachlink
The Valleys Bus Co
Wellington City Transport
Western Buses
  AA Buses
  A1 Service
Yellow Bus Co (New Zealand)
  Whenuapai Bus Travel

The Island Line
Porterbrook Leasing
Sheffield Supertram
South West Trains
Virgin Rail (49%)

Stagecoach was one of the leaders in introducing low-floor double-deckers to London, and in doing so switched from its traditional supplier, Volvo, to Dennis. Some 200 Tridents were in service or on order for Stagecoach operations in London by the end of 1999.

The ubiquitous Dennis Dart SLF was introduced to the operations of Stagecoach Portugal in 1999. The bodywork was by Camo.

Typifying the 1999 intake of single-deckers for Stagecoach is an Alexander-bodied MAN operating in Sunderland for Busways. In 1999 MAN was the main supplier of single-deckers to Stagecoach for its operations outside London. In the capital the group specified the more compact Dennis Dart.

Double-deck coaches reappeared on London to Oxford services (remember City of Oxford's Metroliners in the 1980s?) in the shape of Jonckheere-bodied MANs operated by Stagecoach Oxford. The coloured swoops on the side are a highly-stylised version of the group's corporate red, orange and blue stripes.

Stagecoach expanded in a big way in North America, with its purchase of Coach USA. Typical of modern American coaches is this 45ft-long tri-axle Van Hool 57-seat integral in Florida. It has a Cummins M11 engine and Allison automatic gearbox.

New range from Plaxton

ALMOST HALF the new coaches sold in Britain are Plaxtons, which makes the appearance of a new body from the Scarborough-based manufacturer a major event.

And for coach operators the big news of 1999 was the launch of two new Plaxton models, the Panther and Paragon, at Coach & Bus 99. They will initially be built alongside the existing Excalibur, Premiere and Prima which first appeared at Coach & Bus 91. These are now designated the classic range.

For the early part of 2000 the Paragon and Panther - styled the executive range - will only be available on the Volvo B10M. Later in 2000 they will also be built on the new Dennis R-series (see page 36).

Initially the new models, developed under the Coach 2000 code name, are all being built to a height of 3.5m, although a 3.2m version is likely to follow when the classic range is finally phased out. The timing of that phase-out will depend on the market. For as long as operators want to buy the classic models, Plaxton say they will build them.

The new range offers a gentler style than the old. It has more curves, deeper windows, and a swept-down waistrail which echoes a gentle slope to the front section of the floor which creates the effect of tiered seating for passengers in the first few rows of seats.

There's also some subtle use of chrome-effect trim on the front and rear, capturing some of the retro-quality styling currently favoured by car makers.

Yet the real advances are under the skin. Plaxton has opted for a stainless steel structure, designed to remove concerns about corrosion. Stainless steel was last used in Britain on the Duple 425, and those still in service are reckoned to be remarkably free of rust. The other big advance is in the electrics, with the adoption of multiplex wiring which not only simplifies the wiring but incorporates a diagnostic capability to ease fault-finding.

When the previous range was launched, the Premiere and Excalibur not only looked different, but they had different interior specifications too, with the Excalibur being positioned as an up-market model. That distinction has gone, and the only difference between the Panther and the Paragon is in the design of the front end. Both are available with a choice of interior specifications and Plaxton offers three standard packages: Gold, Platinum and Diamond in order of increasing luxury.

The new range got off to a flying start which seems to have caught even Plaxton by surprise and by early November, just weeks after the launch of the new range, the company had received its 100th order - and was also still taking substantial orders for the older, classic, models.

web sites:
www.plaxton.co.uk
www.henlys.com

One of Plaxton's pre-production Paragons, on Volvo B10M chassis. Deliveries start in 2000.

Volvo to buy Scania

CONSOLIDATION HAS BEEN a constant process in vehicle manufacturing. In the UK alone we have seen Alexander and Dennis united under Mayflower, while in Europe the bus interests of Iveco and Renault are now merged in Irisbus.

But the big one was announced in August: the planned acquisition of Scania by Volvo. The first move was made in January, when Volvo purchased 12.9 per cent of Scania as a prelude to its take-over bid. The sale in March of Volvo's car division to Ford gave the Swedish truck and bus builder the finance with which to pursue its objective. Initially Scania resisted, saying that a Volvo take-over would result in job losses - but in the end it bowed to the inevitable.

For both companies buses represent a small part of their businesses. Volvo produces around 80,000 trucks a year and Scania builds some 40,000. That compares with (in 1998) 10,200 buses and coaches from Volvo and 4,200 from Scania - a little over 10 per cent of their output for both manufacturers.

In the bus and coach market both Swedish builders are strong in Europe and South America. Both build buses in Brazil, and Volvo also has a presence in North America through Prevost and NovaBus, and in Mexico through MASA. In the first two of these operations Volvo is in partnership with Henlys, owner of Plaxton. Volvo has, incidentally, a 10 per cent stake in Henlys which has prompted frequent speculation that it is poised to mount a take-over bid for the company.

In the truck market the two companies compete head-on. But in buses it's rather different. What they do have in common is the exclusive use of their own engines - all Scanias are Scania-powered, and equally all Volvos are Volvo-powered. In both cases high-volume truck engines are adapted for use in the relatively low-volume bus and coach business.

As the accompanying table shows, Volvo's range in the UK is the more comprehensive, ranging from the B6BLE midi through to the versatile new B7L. The mid-engined B10M has long been Britain's best-selling coach, although a replacement is known to be under development.

By contrast Scania has models all aimed at the 12m-long single-deck market: the L94 bus and coach; the heavy-duty K124 coach and the transverse-engined N94 bus which has not yet been seen in the UK. The demise of the N113 looked likely to signal the end of Scania as a builder of double-deckers even before the Volvo take-over.

Volvo has always been well ahead of Scania in terms of UK sales, and this lead was underpinned when Volvo bought Leyland Bus in 1988. However its market leadership has been eroded, thanks in part to the growing successes of Dennis and also to the vagaries of the big bus groups' buying policies. Stagecoach has largely forsaken Volvo for Dennis and MAN; FirstGroup and Travel West Midlands have been buying significant numbers of Mercedes, although TWM has now returned to Volvo for double-deckers. FirstGroup also buys Scanias.

In any take-over of this nature the buyer says that it will maintain the individuality of the acquired brand, and in the short term this will be true. There are parallels in the car industry: BMW and Rover, or Ford and Jaguar, for example.

So don't expect to see any immediate changes in either manufacturer's model line-up. But in the longer term there is bound to be some integration when it comes to major developments, such as a new range of engines.

As the year closed, the take-over was being examined by the European Commission - so it could all come to a halt, which would leave Scania vulnerable to a bid from another vehicle manufacturer.

• In November Volvo purchased a 19.9 per cent stake in the truck and bus division of Japanese manufacturer Mitsubishi.

websites:
www.scania.com
www.bus.volvo.se
www.volvobuses.volvo.co.uk

VOLVO AND SCANIA IN THE UK: 1999 DELIVERIES

	Bus	Coach
Volvo	462	421
Scania	140	149

VOLVO AND SCANIA IN THE UK: YEAR 2000 RANGE

Model	Type	Engine size/location
Bus		
Volvo B6BLE	10.5m midi	5.48 litre/rear
Volvo B10BLE	12m sd	9.6 litre/rear
Scania L94UB	12m sd	9 litre/rear
Volvo B7L	12m sd	7.3 litre/rear
Volvo B10LA	18m artic	9.6 litre/rear
Volvo B7LA	18m artic	7.3 litre/rear
Volvo B7L	9.5m dd	7.3 litre/rear
The B10BLE and B10LA are being phased out in favour of the B7L and B7LA.		
Coach		
Volvo B9M	10m	9.6 litre/mid
Volvo B7R	12m	6.7 litre/rear
Scania L94IB	12m	9 litre/rear
Volvo B10M	12m	9.6 litre/mid
Volvo B10MT	12m three-axle	9.6 litre/rear
Scania K124IB	12m	12 litre/rear
Scania K124IB	12m three-axle	12 litre/rear
Volvo B12T	12m three-axle	12.1 litre/rear

New in the South

The Oxford bus company took 15 Wright-bodied Volvo B10BLEs in 1999. They joined Wright-bodied Darts which were delivered in 1998. Oxford is unusual among operators outside London in specifying two-door bodywork.

The first new service bus for Worths of Enstone for many years entered service in the spring. It was a Dennis Dart SLF with Plaxton Pointer 2 bodywork.

Redwing of London was 1999's biggest UK buyer of the Iveco EuroRider, taking 14 with Beulas Stergo E bodywork. One was entered at the Brighton Coach Rally.

Cambridge Coach Services was sold to National Express in 1999. Its last new coaches under Blazefield control were Plaxton-bodied Volvos, delivered at the start of the year. This Premiere was equipped with a wheelchair lift and was replacing the Van Hool-bodied B10M seen in the background.

Brighton & Hove took 20 Dennis Tridents in 1999, with another 20 to follow in 2000. They had East Lancs bodies. East Lancs is an established supplier of double-deckers to the company, but in the past they have been on Scania chassis.

New 'deckers for London fleets

THE BIG CHANGE in London in 1999 was the arrival of the low-floor double-decker. At the start of the year there were around two dozen running for Arriva. By the end of the year that number was over 600, running for Arriva, Blue Triangle, Capital Logistics, FirstGroup, Metrobus, Metroline and Stagecoach. The buses for Arriva and Capital Logistics were DAFs; all other operators were buying Tridents. The first low-floor Volvo double-deckers will not appear until 2000.

Arguments about the internal layout of low-floor double-deckers continued, as operators and London Transport Buses weighed up the pros and cons of improved access against reduced lower-deck seating capacities. Lessons being learned will no doubt have a bearing on the interior design of future double-deck orders.

The last old-style double-deckers for London entered service with CentreWest in the spring. These were 20 Volvo Olympians with two-door Northern Counties Palatine I bodies which were part of a speculative build of 22 London-specification double-deckers by the manufacturers. The other two went to Ireland.

Tendering continued to bring change and colour to London's streets - and in some cases rather more colour than was intended.

The route 60 saga in south London continued to unfold. Capital Logistics should have taken over the service, running from Streatham to Old Coulsdon, from Arriva London South in August 1998. To do so it had ordered 16 DAF double-deckers - 10 with Plaxton President bodies and six Optare Spectras. The new low-floor 'deckers were not ready in time for the start of the service, and Capital - allegedly finding driver recruitment difficult - subcontracted the operation to Stagecoach Selkent and Blue Triangle. It was then agreed that a new operator, Driver Express of Horsham, would take over the 60, along with the new vehicles being built for it.

Driver Express, trading as Omnibus London, should have taken control of the route in January - but didn't. Instead it reverted to Capital Logistics, and a range of hired vehicles appeared until, eventually, the new vehicles trickled through in the spring. There was a time lag between the DAFs being bodied and entering service, suggesting a reluctance on the part of the manufacturers to part with their valuable buses until they were sure of payment from troubled Capital Logistics. In May Capital Logistics was bought by Tellings Golden Miller. Driver Express, meanwhile, closed down.

Independent Way, the company which rescued the Limebourne operations when Q-Drive folded in October 1998, started the year running hired buses on the 42 (Liverpool Street Station to Denmark Hill) and 156 (Wimbledon to Clapham Junction). These were Arriva (aka Grey Green) Volvo B10Ms and London General Marshall Minis. New Dennis Darts with SC Coachbuilders bodies replaced them in the spring.

In August Limebourne took responsibility for the G1 (Battersea to Streatham) as a sub-contractor to Arriva London South which was short of drivers. It operated the service using 11 Darts supplied by Arriva. Limebourne also won the 344 from London General, for which it bought more SC-bodied Darts.

Armchair expanded, taking a fleet of 13 new Dennis Darts with Alexander bodies for the 209, which crosses Hammersmith Bridge. However a temporary weight restriction on the bridge saw the 209 stay with previous operator London United, while Armchair temporarily took on the 49, which London United was scheduled to take over from London General. In November the weight restriction was lifted and the routes switched to their intended operators, Armchair on the 209 and London United on the 49.

In April a new name appeared on a London bus service - old-established coach operator Mitcham Belle. It took over the 127 from London General and Blue Triangle - the latter had replaced Selkent in the early part of the year to release Selkent's resources for new routes in Bexleyheath. Mitcham Belle bought 12 Plaxton-bodied Darts for the service.

Metroline opted for Dennis Tridents with Alexander and, as here, Plaxton bodies. Route 43 received a major upgrade with new buses and improved bus priorities.

Another new name was London Traveller, a company which operated a variety of contracts but no scheduled routes in London. It won the 187, initially running with an assortment of hired low-floor buses from Dennis, MAN and Volvo until all of its 15 new Volvo B6BLEs with East Lancs bodies were delivered. Most of these were pressed into service in all-over red. London Traveller was reformed in the summer, and became the trading name of the new Metropolitan Omnibus Co, in which Yorkshire Traction had a 25 per cent share. Unusually for a London bus operation it had an address in Devon.

Blue Triangle, which has long provided rail replacement buses in London and has operated commercially in Essex, won the 474 which it started with older buses - mainly MCW Metrobuses - although these were replaced in the autumn with nine new Dennis Tridents bodied by East Lancs.

Smallest of the newcomers in all senses of the word was Hackney Community Transport which in June took over the Enfield and Hackney mobility network from Arriva London North using three minibuses.

Metrobus of Orpington also took Tridents with East Lancs bodies - the company's first new Dennis double-deckers. They entered service on the 161, running from North Greenwich to Chiselhurst. Soon after, in September, the Go-Ahead Group bought the company. With 200 buses Metrobus had been the largest independent serving London. It now joins London Central and London General in Go-Ahead ownership, but is to retain its own identity.

A rather unhappier fate awaited Harris Bus in December. After some speculation that the operation was likely to be bought by FirstGroup, there came the surprise announcement that the receivers had been called in. Harris Bus operated 76 buses on London Transport contracts, and at the end of the year the receivers were running the company while looking for potential buyers. Harris Bus had been operating LT tendered services since April 1997 and the speed of its growth had been quite remarkable. As was the speed of its demise.

The closure in May of Arriva Croydon & North Surrey's Leatherhead garage saw one route - the 465 - being transferred to Tellings Golden Miller, along with the seven Dennis Darts which operated it. A second route, the 85, went to London United along with 12 modern DAF double-deckers. Shortly after Arriva Croydon & North Surrey also closed its Hounslow depot, with the routes and 14 buses passing to Tellings Golden Miller. Further retrenchment at Arriva Croydon & North Surrey came in October, when the former Londonlinks Croydon depot with its seven routes (289, 315, 367, 403, 407, 455, 627) and the 62 buses operating them were transferred to Arriva London South.

Major changes in Harrow at the end of the summer saw Sovereign Buses (London) - as Sovereign Buses (Harrow) had become in May - introduce 57 new Dennis Darts to routes in the area, ousting its own Mercedes minibuses and also replacing Metroline on two routes. Sovereign Buses (London), absorbed the London Sovereign fleet in August. London Sovereign was the successor to BTS coaches and the bright red livery worn by its Olympians is to be phased out in favour of Sovereign blue and cream. Its Routemasters on the 13 (Aldwych to Golders Green) remain red.

Arriva the Shires gained a short route in Harrow, the H18, to which it introduced Wright-bodied Volvo B6BLEs.

In Uxbridge, coach operator Wings established a bus operating arm when it took over the U7, with three East Lancs-bodied Darts, in November. A further four Darts were ordered for a new Uxbridge route to start in 2000.

The Croydon Tramlink was due to start in November, which was then postponed to December. In the event, no trams were carrying passengers in 1999, although they could be seen on trial runs. When the service does commence there will be major route revisions in which Metrobus and CentreWest win out at the expense of Arriva London South.

To match the modern imagery of the Millennium Dome at Greenwich, open from January 2000, London Transport awarded the tender for the Millennium Transit services M1 and M2 to London Central. Part of the route is on a short - 1.3 kilometre - two-lane busway with drive-by-wire electronic guidance. London

Sovereign Buses (London) upgraded its Harrow network with new Dennis Darts, taking 57 with Plaxton bodywork. Some were for new routes; others replaced Mercedes minis on existing routes.

Next stop unclear. Harris Bus looked alive and well when the Millennium Dome was under construction in the summer, but had called in the receivers before the year ended. An Optare Excel is seen at the impressive new North Greenwich bus station, with the Dome in the background.

FirstGroup's CentreWest and Capital Citybus fleets both received Dennis Tridents with Plaxton bodies. Both fleets used this revised livery with upswept bands of yellow and white relief on the lower deck side panels.

Central ordered 17 DAF SB220s fitted with a new style of East Lancs body - the Millennyum - two of which were at Coach & Bus 99. Three of the DAFs will run on LPG. The routes from Charlton Station (M1) and Greenwich Station (M2) will serve the impressive new bus station at North Greenwich, which opened in May.

Among tender awards to be implemented in 2000, the big surprise was the award of two routes to Connex, the French-owned operator of rail services. The first was the 3 (Oxford Circus to Crystal Palace), which is currently operated by London Central using Optare Spectras, and for this Connex ordered 27 Alexander-bodied Tridents. At the end of the year it was announced that it had also gained the 322 (Crystal Palace to the Elephant & Castle). The 3 is to be taken over by Connex in February 2000; the 322 in April.

Aside from the low-floor double-deckers mentioned above, most new buses for London operation were Dennis Darts. Among the big names, new Darts were delivered to FirstGroup, Go-Ahead, London United, Metroline and Stagecoach. The combined success of the Dart and the Trident meant that new Dennises easily outnumbered new buses from DAF and Volvo in the capital.

One type to disappear during 1999 was the Renault S75, with the last of CentreWest's big fleet being sold or transferred to other FirstGroup operations during the

London Traveller expanded in 1999, adding six Volvo B6BLEs. Some entered service in unrelieved red. They had East Lancs Spryte bodies.

Route 60 entered the annals of London bus history as a series of crises affected its operation. Blue Triangle was one of many operators which stepped in to provide vehicles while Capital Logistics and Driver Express stumbled along. This is a former London Transport Metrobus.

Eventually new low-floor buses did appear on the 60, under the ownership of Capital Logistics which was itself about to be taken over by Tellings Golden Miller. The new fleet was made up of DAF DB250s bodied by Optare and Plaxton.

Tellings Golden Miller took over a number of services from Arriva Croydon & North Surrey, an operation which was significantly cut back in 1999. These included services in Hounslow, along with the buses operating them. A former Arriva Dart, still in London & Country colours, carries the TGM name on the side. The body is by East Lancs.

Metrobus took 15 Dennis Tridents with East Lancs Lolyne bodies for use on a service running to the site of the Millennium Dome at Greenwich. Metrobus was bought by the Go-Ahead group in September.

Airlinks the Airport Coach Company was the highly-descriptive if less than catchy new name for Speedlink. Airlinks put new buses onto the H30. These were Dennis Darts with air-conditioned Plaxton bodies.

year. The Renaults had clocked up nine years of service. The Marshall Minibus almost vanished. Although barely three years old, those operated by CentreWest and London General were returned to Marshall - quite an indictment of the vehicles' problems. A sole survivor in the Metroline fleet - ex–MTL London - still runs in Potters Bar.

And, less anybody think that the vast influx of new buses sounded the death knell for the Routemaster, two contracts were announced in 1999 which ensure the type's survival until 2004 on the 19 (Finsbury Park to Battersea) and 36 (Queens Park Station to Lewisham) with Arriva London South and London Central respectively.

A simplified fares structure for London's buses was announced in November for implementation in January 2000. Two fares - £1 in zone 1 and 70p elsewhere- replace graduated fares ranging from 60p to £1.20.

website:
www.londontransport.co.uk

MAJOR CHANGES IN TENDERED ROUTES 1999

25	Oxford Circus - Ilford	East London to First Capital
49	Clapham Junc - Shepherds Bush	London General - London United
85	Kingston - Putney	Arriva Croydon & North Surrey to London United
96	Woolwich - Dartford	Arriva Kent Thameside to Selkent
99	Woolwich - Erith	London Central to Selkent
127	Purley - Tooting	London General/Blue Triangle to Mitcham Belle
140	Heathrow-Harrow Weald	London United to Metroline
178	Woolwich - Kidbrooke Stn	London Central to Selkent
183	Golders Green - Pinner	Metroline to Sovereign (London)
187	South Harrow - Queens Park	Metroline to London Traveller
196	Norwood Junction - Brixton	London Central to London General
209	Hammersmith - Mortlake	London United to Armchair
258	Watford Junction - South Harrow	First Beeline to First CentreWest
269	Bexleyheath - Bromley North	Arriva Kent Thameside to Selkent
291	Woolwich - Plumstead	London Central to Selkent
337	Clapham Junc - Richmond	London General to London United
344	Clapham Junc - London Bridge	London General to Limebourne
422	Bexleyheath - N Greenwich	Arriva Kent Thameside to London Central
465	Teddington - Dorking	Arriva Croydon & North Surrey to TGM
474	Canning Town - E Beckton	Blue Triangle
492	Sidcup - Bluewater	Arriva Kent Thameside to London Central
601	Thamesmead - Dartford Heath	London Central to Arriva Kent Thameside
B14	Bexleyheath - Sidcup	Arriva Kent Thameside (484) to Crystals
B15	Bexleyheath - Joydens Wood	Arriva Kent Thameside to London Central
D7	Mile End Station - Poplar	East London to First Capital
H12	South Harrow - Stanmore	Metroline to Sovereign (London)
H18	Harrow Weald - Harrow	Metroline to Arriva the Shires
H20	Hounslow - Ivybridge	London United to TGM
U7	Uxbridge Station - Charville Lane	First CentreWest (207A) to Wings

Arriva expands, but mainly in Europe

With opportunities for significant UK expansion on the wane, Arriva, like other big British bus groups, looked overseas for potential acquisitions in 1999. Thus in March it strengthened its position in Europe with the purchase of the 777-strong Bus Danmark fleet. This was followed in July by Spanish operator Ideal Auto, running 175 vehicles. A second, smaller, Spanish business, Finisterre Transporte, was purchased in September. By this time Arriva's European fleet was approaching the 3,000 mark - with around 1,500 buses in Holland, 1,000 in Denmark, 230 in Spain and 200 in Sweden. This compares with 9,000 vehicles in Britain.

In October the group was in discussions to secure the purchase the 830-vehicle Norgessbus business in Norway, but these came to nothing and were abandoned at the end of the year.

At home Arriva Yorkshire West took over the routes of Cygnet Travel of Darton when that company closed in February. No vehicles were involved. In August neighbouring Arriva Yorkshire North took over the bus operations of Jaronda of Selby, acquiring one of Jaronda's vehicles in the process, and in December a rather bigger Yorkshire take-over happened with the acquisition of Ambermile, which traded as K-Line in Huddersfield and as Taylors in Leeds. This involved around 40 buses.

In August, Arriva Midlands North cut back in Burslem, with its operations there - and 20 vehicles - passing to Crystal Coaches, a new operator. This was followed in November by the sale of 20 coaches which were operated under the Stevensons name by Arriva Midlands North. These went to rapidly-expanding Dunn-Line of Nottingham. But December saw expansion in the north Midlands area when Arriva North West bought the 40-bus Nova-Scotia Travel of Winsford, Cheshire.

Eden Bus Services, acquired by United Auto in October 1995, lost its separate identity in the summer. Its operations were absorbed by Arriva Durham County in July.

New vehicles for Arriva in 1999 were quite a mixture. The only double-deckers outside London were 18 old-style step-entrance Optare Spectras on DAF DB250 chassis for Arriva Yorkshire. The group's London fleets received some 150 DAF DB250 low-floor 'deckers. The vast majority had Alexander ALX400 bodies, but there were 20 Plaxton Presidents. Two of the Alexander-bodied DB250s were sent on long-term loan to Dublin Bus for evaluation.

Single-deckers were varied. There were Plaxton Pointers on Dennis Dart chassis for a wide range of Arriva companies, and Plaxton Prestiges on DAF SB220s for a few. Wright supplied bodywork on Volvo B6BLEs and B10BLEs - Volvos went to Arriva the Shires, Arriva Herts & Essex, Arriva Northumbria and Arriva Scotland West. Nine SB220s with Ikarus bodies were bought by Arriva North East. Alexander-bodied Darts entered service with Arriva London.

Of particular interest were 17 DAF SB220s with LPG engines - four for Arriva Cymru, six for Arriva Scotland West and seven for Arriva the Shires. All had Plaxton's Prestige bodywork.

A small number of new DAF coaches entered service in 1999, with bodywork by Plaxton and Van Hool. The Plaxtons were for inter-urban services in north-east England.

Orders announced in 1999 for delivery in 2000 included more Dennis Darts with bodywork by Alexander and Plaxton; the group's first Dennis Tridents, breaking DAF's exclusive hold on Arriva double-deck orders; and full-size single-deck Alexander ALX300s on both DAF and Volvo chassis. An order was also placed for 20 Volvo B7L double-deckers with Alexander bodies for use on services operating on the Leeds guided busway. They will be the world's only guided double-deckers - although they're not the first: the West Midlands PTE briefly experimented with guided Metrobuses in the 1980s. And while no specific announcement was made, there will be Wright-bodied DAF SB120s, too.

Arriva's livery may be standardised, but there's certainly variety in its new bus intake.

website: www.arriva.co.uk

ARRIVA COMPANIES	
Arriva Passenger Services	Arriva Scotland West
Arriva Cheshire East	Arriva the Shires
Arriva Colchester	Arriva Southend
Arriva Crawley & East Surrey	Arriva Southern Counties
Arriva Croydon & North Surrey	Arriva Surrey & West Sussex
Arriva Cymru	Arriva Tees & District
Arriva Derby	Arriva Teesside
Arriva Durham County	Arriva West Sussex
Arriva East Herts & Essex	Arriva Yorkshire
Arriva Fox County	Arriva Yorkshire North
Arriva Guildford & West Surrey	Arriva Yorkshire South
Arriva Kent & Sussex	Arriva Yorkshire West
Arriva Kent Thameside	Leaside Travel
Arriva London North	New Enterprise Coaches
Arriva London North East	The Original London Sightseeing Tour
Arriva London South	
Arriva Manchester	
Arriva Merseyside	
Arriva Midlands North	Arriva Danmark
Arriva North East	Arriva Nederland
Arriva North West	Arriva Noroeste Espana
Arriva Northumbria	Arriva Sverige

Plaxton Prestige production drew to a close in 1999 with the last deliveries including 17 gas-powered DAFs for Arriva. These included Scotland's only LPG-fuelled buses, bought by Arriva Scotland West for use on the Scottish Citylink service connecting Glasgow city centre with the airport.

Unusual additions to the Arriva North East fleet were nine DAF SB220s with Hungarian-built Ikarus bodies.

Low-floor DAF double-deckers entered service with Arriva in London and included 20 with Plaxton President bodywork.

Arriva Yorkshire took a batch of step-entrance Optare Spectras in the summer. They were the last for the group. All other new double-deckers for Arriva in 1999 were of low-floor layout. *David Barrow*

Nova-Scotia sold out to Arriva in December. Earlier in the year it had added five new Dennises to its fleet, four of which were Plaxton Mini Pointer Darts.

Double-deck revival

DELIVERIES OF NEW double-deck buses to British operators in 1999 were up by almost 25 per cent on 1998 from 756 to 933, the highest figure since 1989. The remarkable growth reflects the arrival in volume of the new generation of low-floor models. And in this race there are clear winners and losers.

The out-and-out winners have to be Dennis. In 1999 they delivered just over 600 Tridents to UK fleets, which compares with a best-ever figure for the old Dominator model of just 159 in 1982. Add to that outstanding orders for delivery in 2000 and there will soon be approaching 1,000 Tridents in UK service - in just two years. Total production of two-axle Dominators only just exceeded that figure.

And the manufacturer who was the first with a low-floor double-decker in revenue-earning service, Optare, seems to be in danger of dropping out of the race. It supplied around a dozen low-floor Spectras in 1998, and that rose to almost 30 in 1999 - but with no orders announced for 2000.

There are currently three low-floor double-deck chassis in operation in Britain from - in descending order of deliveries in 1999 - Dennis, DAF and Volvo. A fourth manufacturer, MAN, is looking at the market.

Dennis officially launched its Trident at the start of the year, bringing four buses together to do so which showed the varying approaches to low-floor 'deckers being adopted by different operators and bodybuilders. The table on the opposite page shows key data.

With the Trident Dennis has made some remarkable conquests. Having standardised on the Olympian since it started buying new double-deckers back in 1988, Stagecoach has become a major Trident user and runs almost 200 in London, Manchester, Newcastle and Grimsby-Cleethorpes. Another fleet where the idea of Dennis double-deckers would have been unimaginable just three years ago is Lothian, which has 95 in service or on order.

And the list goes on - from Metrobus to Metroline, or from Brighton & Hove to Preston Bus. The first low-floor double-deckers in Wales, running for Newport Transport, are Tridents. Tridents are appearing all over the country in fleets which would never have considered a Dennis double-decker in the past.

DAF sits in the number two position thanks mainly to orders from Arriva, and all of these are operating in London which has the country's biggest concentration of low-floor 'deckers. However there have been other orders for low-floor DAFs, most notably from Travel West Midlands - although that operator has since switched to Volvos. A few stock-build DAFs have gone to small operators: A-Bus in Bristol and UK North in Manchester.

And the reason Volvo is trailing? That's down to the time it has taken the company to develop a new model. The original B7, showed under Plaxton's new President body at Coach & Bus 97, had to be extensively reworked to make it acceptable to British operators. The latest version clearly is, with an order for 100 having been placed by Travel West Midlands, and with other large orders coming from Dublin Bus, FirstGroup, Go-Ahead and London United. At its July launch Volvo claimed over 450 orders for its B7 'decker, which it later showed at Coach & Bus 99 where it was billed - perhaps optimistically in the light of current patterns of orders - as the natural successor to the market-leading Olympian.

The three bodybuilders who are offering to body all three chassis - Alexander, East Lancs and Plaxton - have stylish new products. During 1999 Plaxton's President was supplied on all three chassis and to new customers too, such as Lothian, Metroline and Travel West Midlands which had not previously bought double-deckers from Plaxton's Wigan plant, known until the start of 1999 as Northern Counties.

Alexander's attractive ALX400 was the favourite at Arriva and Stagecoach, on DAF and Dennis respectively. And orders for the neat East Lancs Lolyne came primarily from operators with previous experience of East Lancs' products, such as Brighton & Hove, Metrobus, Nottingham City Transport and Yellow Buses in Bournemouth. All 1999 Lolynes were built on Dennis chassis.

Optare, building exclusively on the DAF DB250, delivered low-floor Spectras to Travel West Midlands and Wilts & Dorset in 1999.

Small numbers of Volvo Olympians entered service during 1999 and the last will be delivered in 2000. Bullock of Cheadle took four, the last to have Northern Counties Palatine II bodies.

However, like old soldiers, the previous generation of double-deckers is fading away rather than suddenly dying. The last Northern Counties Palatine bodies were completed in the spring, all on Volvo Olympian chassis. These were four Palatine IIs for Bullock of Cheadle, followed by 22 stock build Palatine Is of which 20 went to CentreWest and the remaining two to Kavanagh of Urlingford.

Alexander's Falkirk factory switched from building R-types to ALX400s, and the last British R-types went to Stagecoach. However Alexander (Belfast) continued building R-types for Dublin Bus, and has an order for seven Royales for Yorkshire Coastliner, to be delivered in 2000.

East Lancs continues to produce small numbers of Pyoneers on Olympians, with the last due in 2000. And Scania still has stock Cityzens to sell, on N113 chassis. When these go, they look set to spell the end of Scania as a builder of double-deck chassis. It has been supplying UK operators with N-series chassis for double-deck bodywork since 1980.

Among the last old-style step-entrance DAF DB250s were 18 Optare Spectras for Arriva Yorkshire.

All of this means that the end is nigh for Britain's best-selling double-decker of the last two decades: the Olympian, the 10,000th example of which was handed over to Dublin Bus in the autumn. And with it ends the dominance of the double-deck market which Volvo inherited from Leyland. The B7 will recover some of the ground which Volvo lost to Dennis in 1999 - but it will be struggling to narrow the wide lead which Dennis has established.

UK OPERATORS OF LOW-FLOOR DOUBLE-DECKERS

Operator	Make	Operator	Make
A-Bus, Bristol	DAF	Nottingham City Transport	Dennis
Arriva London	DAF	Oxford Bus Company	Dennis
Blue Triangle	Dennis	Preston Bus	Dennis
Brighton & Hove	Dennis	Solent Blue Line	Dennis
Bullocks, Cheadle	DAF	Stagecoach Busways	Dennis
Capital Logistics	DAF	Stagecoach East London	Dennis
First Bristol CityLine	Dennis	Stagecoach Grimsby-Cleethorpes	Dennis
First Capital	Dennis	Stagecoach Manchester	Dennis
First CentreWest	Dennis	Travel West Midlands	DAF/Volvo
Lothian	Dennis	UK North, Manchester	DAF
Mayne, Manchester	Dennis	Wilts & Dorset	DAF
Metrobus	Dennis	Yellow Buses, Bournemouth	Dennis
Metroline	Dennis		
Newport Transport	Dennis		

LAUNCHING THE TRIDENT: FOUR MODELS COMPARED

Operator	Bodybuilder	Passenger capacity	Unladen weight (kg)
Brighton & Hove	East Lancs	97 (47/30 + 20)	11620
East London	Alexander	100 (51/23 + 26)	11520
Metroline	Plaxton	95 (45/22 + 28)	11660
Nottingham	East Lancs	103 (53/35 + 15)	11240

In corporate colours for the Coach & Bus show, but due to be repainted in park-and-ride livery, was this First City Line Dennis Trident with East Lancs Lolyne body. The optional bonded glazing enhances the Lolyne's looks.

New in the North East and Yorkshire

Moor-Dale became part of the new Status group in 1999. New coaches during the year included Volvo B10Ms with Plaxton Premiere 350 bodywork.

Wright's Renown body was a popular choice on the Volvo B10BLE in 1999. Keighley & District added a repeat order for 12 to its fleet.

Scarborough & District supported its local coachbuilder, putting five new Plaxton-bodied Dennis Dart SLFs into service in 1999. The lettering on the side - Low Floor, Nice & Easy - is also used by other EYMS companies running low-floor buses.

Scarlet Band is one of the few survivors from the large number of independents which once served the north-east of England. Its first low-floor buses were three Mini Pointer Darts.

Plaxton's Cheetah appeared in many small fleets during 1999, including Tynedale Group Travel of Haltwhistle. The Cheetah is built only on the Mercedes-Benz Vario chassis.

Powells Bus of Rotherham added two Mini Pointer Darts to its fleet, for use on rural services. The Mini Pointer Dart with its step-free entrance found its way into many small fleets which in the past would have been more likely to have bought Mercedes-Benz minibuses.

More growth at FirstGroup

THE MAJOR UK FirstGroup news in 1999 was the acquisition in April of one of the few former NBC companies which had remained in the ownership of its original management buy - out team. This was Cawlett, purchased for £10.6million.

Cawlett - better known through its main subsidiaries North Devon Red Bus and Southern National - operated over 400 buses. But they were on the whole rather elderly buses and the average age of the fleet was reckoned to be a high 11 years. The Cawlett companies ran substantial numbers of first-generation Ford Transit minibuses, a type long since vanished from virtually every other major fleet. They also operated Atlanteans, VRTs and Leopards - but had also been investing in new Mercedes Varios and Dennis Darts.

The last new buses delivered prior to the take-over were four Plaxton-bodied Darts for the company's West Dorset fleet. And the first new vehicles after the take-over were six Darts with East Lancs bodies for Southern National. These were the first East Lancs bodied-buses to carry FirstGroup corporate colours. A tidying up of the former Cawlett group's operations saw Smiths of Portland close in September, while to expedite fleet modernisation nine new Volvo B6BLEs were diverted from First Bristol to West Dorset Coaches.

Control of Southern National and its associates passed to First Bristol, while North Devon's operations fell under the control of First Western National. The take-over confirms FirstGroup as the major operator from Bristol to Land's End, the only other big operators in the south-west being Stagecoach in and around Exeter, and municipally-owned Plymouth Citybus. FirstGroup is also the major operator of trains in the region, through Great Western Trains.

First Bristol, incidentally, annoyed the traffic commissioner in July. When it was banned from operating between Bath and Corsham following concerns about running unregistered journeys, sister company First Wessex immediately registered the service. Banning a company from operating seems more likely to inflict punishment on bus users than on bus operators, but simply switching a service registration from one group company to another was seen by the traffic commissioner as circumventing his powers of sanction.

A smaller take-over in July - involving just 25 vehicles - saw FirstGroup buy Biss Brothers and Airport Coaches from Lynton Travel. Biss was a coach operator with a long history, originally based in Bishops Stortford but more recently at Stansted Airport. Seven coaches and the Biss name were sold to Associated Coachways of Harlow in September, and the remaining contract operations were absorbed by Essex Buses through a new Airport Buses company. Smaller still was the acquisition by First Quickstep of the Optional Bus operation in Leeds, along with six Atlanteans. Optional had been running for two years in a two-tone green and red livery modelled on that designed by Best Impressions for London & Country.

In Scotland, FirstGroup was in June named as the preferred bidder to build and operate CERT - City of Edinburgh Rapid Transit - on a 25 year franchise. CERT plans a rapid transit link between the city centre and the airport using articulated buses on a guided busway. It is intended to start in 2001.

Also in June, First Edinburgh was adopted as the trading name for Midland Bluebird and Lowland Omnibuses, replacing both these names, along with the historic SMT name in use in the Edinburgh area. The First Edinburgh name covers an area from Berwick on the English border, to Balfron, 15 miles north of Glasgow - which must be a shade baffling for travellers in places so far from the Scottish capital. However geographically inappropriate it may be, the First Edinburgh name is carried by 600 vehicles. The Fife First identity remained unchanged.

First Glasgow had a major relaunch of its trunk services in August using Overground as a brand name and introducing minimum 10-minute day-time frequencies on 18 radial routes, which were shown on a London Underground-style schematic map. Strangely, the Overground map showed only Overground routes, giving no clue that First Glasgow operated a host of other less-frequent services in and around the city. Did I hear someone say two-tier service?

In Wales, First Cymru expanded following the collapse of old-established independent Davies of Pencader, which went into receivership. The Davies bus operations and 38 vehicles were taken over by First Cymru in June.

Like some other British

Alexander started supplying ALX200 bodies to FirstGroup in 1999. The first ten went to First Calderline in Halifax.

operators, FirstGroup moved into the USA in 1999, with the announcement in July that it was buying the 10,000-vehicle school bus operator Ryder, and the management services company Ryder ATE. This was quickly followed by Bruce Transportation, operating around 1,000 vehicles on school services in New Hampshire, Massachusetts and Vermont.

However don't look for the appearance of FirstGroup's livery across the Atlantic. School buses are yellow, while the fleets run by Ryder ATE are management contracts, typically running a fleet owned by a local transit authority. Ryder ATE was renamed First Transit, while Ryder became First Student.

The most striking new buses for FirstGroup in 1999 were its 40 Volvo B10LA artics with 140 - passenger Wright bodies. These have just 55 seats which means that to achieve maximum loading 85 people have to stand. Hmmm... (But on the plus side, the flat floor bring the benefits of easy access to many more passengers than is possible on a double-decker.) The first 15 entered service in Manchester in April. They were followed by 15 for Leeds and, at the end of the year, 10 for Glasgow.

The artics are 17.84m long, cost around £200,000 each, weigh 16,096kg and are powered by 285bhp Volvo DH10A engines driving through ZF 5HP590 automatic gearboxes. The engine is mounted horizontally at the rear of the trailing section.

Ironically, as this fleet of new articulated buses was entering service, the fleet of 13 Leyland DABs operated by First Mainline came out of service. New to the South Yorkshire PTE in 1985, these were for many years the only articulated buses operating in Britain.

Other new buses were mainly Wright-bodied Scanias and Volvos, which were delivered to subsidiaries in most parts of Britain. The group's first Alexander ALX200s - ten for First Calderline - were delivered in the spring. FirstGroup also bought large numbers of Marshall bodies on Dennis Dart chassis for its London fleets. It was Marshall's biggest customer in 1999. Small numbers of Optare Solos entered service with First PMT and First Manchester.

Most of the double-deckers for FirstGroup in 1999 were Plaxton-bodied Dennis Tridents for operation in London. However there were 10 Alexander-bodied Olympians for Yorkshire Rider and a few for Midland Bluebird, as well as 20 Northern Counties-bodied buses for London which had been part of a speculative build programme by the manufacturers. In addition, 26 East Lancs-bodied Tridents were supplied at the end of the year for park-and-ride operation in Bristol. One of these was at Coach & Bus 99, carrying corporate colours for the show - although it was repainted in a dedicated park-and-ride livery before entering service.

Striking new coaches were delivered to First Beeline in the summer for the Reading to Heathrow rail-air link. These were three-axle Volvo B12s with Plaxton Excalibur bodies - a rare example of Plaxton bodywork on a tri-axle chassis.

Orders announced in July (see table) included more double-deckers, with both Plaxton-bodied Tridents and Alexander-bodied Volvo B7Ls - a mix which sees Mayflower group's chassis getting someone else's bodies, while Mayflower group bodies are going on to another maker's chassis. Small numbers of Mercedes O405s are operated by FirstGroup, and for 2000 it will be taking 60 of the new Citaro model. One was at Coach & Bus 99.

An unusual bus was repatriated from Hong Kong during the year - a New World First Bus Leyland Victory 2, which had been new to China Motor Bus in 1980. The Victory was a rugged front-engined chassis - developed by Guy and powered by Gardner - which had been produced in response to concerns about the poor reliability of rear-engined buses in Hong Kong (and South Africa) in the early 1970s. The former CMB vehicle, with 102-seat Alexander bodywork, was donated to the Scottish Vintage Bus Museum in Fife.

And an import of possibly greater significance was that of an Alexander-bodied Volvo Olympian bought from Singapore Bus Services which could be disposing of some more of its most modern buses without air-conditioning. It was repainted in First PMT livery in the autumn.

FirstGroup runs around 10,000 buses in the UK and some renaming of companies took place during the year to bring more of them into line with the formal style which starts the company name with First. Thus, for example, Bristol Omnibus became First Bristol Buses, Lowland Omnibuses became First Edinburgh, and Western National and the Berks Bucks Bus Co were also given new First-style names.

website: www.firstgroup.com

FIRSTGROUP: 2000 ORDERS

Double-deck	
Volvo B7L/Alexander ALX400	100
Dennis Trident/Plaxton President	130
Full-size single-deck	
Volvo B10BLE/Wright	105
Scania L94/Wright	75
Mercedes-Benz Citaro integral	60
Volvo B7/Wright artic	20
Midi single-deck	
Dennis Dart SLF/Alexander ALX200	100
Dennis Dart SLF/Marshall Capital	80
Volvo B6BLE/Wright Crusader 2	60
Dennis Dart SLF/Plaxton SPD	10
Minibus	
Optare Solo	30
Coach	
Volvo B10M/Plaxton	10

Low-floor Plaxton Pointer 2s for FirstGroup are generally in corporate colours. A trio delivered at the turn of the year was unusual in being in company colours - First Manchester orange. They had been ordered by Timeline Travel, and did not have FirstGroup interior trim - and therefore did not get FirstGroup livery.

The Cawlett group was taken over by FirstGroup in April. Just weeks before the takeover Cawlett's Dorset Transit fleet received four new Dennis Dart SLFs with Plaxton Pointer 2 bodies.

Impressive new coaches for First Beeline were seven tri-axle Volvo B12Ts with Plaxton Excalibur bodies. They were for the Rail-Air link between Heathrow airport and Reading station.

First Wessex took delivery of Volvo B10Ms with Plaxton Expressliner bodies for operation on the National Express Flightlink network.

Few double-deckers were delivered to FirstGroup subsidiaries outside London. They included the last of a batch of 15 for First Midland Bluebird which had Alexander Royale bodies to FirstGroup's high-quality specification.

Articulated buses were introduced to FirstGroup operations in Manchester, Leeds and Glasgow in 1999. First Manchester received 15 of these Volvo B10LAs with Wright Fusion bodies. *David Barrow*

FIRSTGROUP SUBSIDIARIES AND PRINCIPAL TRADING NAMES

- Capital Citybus
  - First Capital
- CentreWest London Buses
  - First Challenger
  - First Ealing Buses
  - First Gold Arrow
  - First London Buslines
  - First Orpington Buses
  - First Uxbridge Buses
  - Heathrow Fast
- Essex Buses
  - Airport Buses
  - First Eastern National
  - First Thamesway
- First Aberdeen
  - Kirkpatrick of Deeside
  - Mairs of Aberdeen
- First Beeline
- First Bristol Buses
  - First Badgerline
  - First CityLine
  - First Durbins
  - First Southern National
    - First Dorset Transit
  - First Streamline
- First Cymru Buses
  - First Brewers
  - First SWT
- First Eastern Counties Omnibus
  - First Blue Bus
  - Halesworth Transit
    - Flying Banana
  - Rosemary Coaches
- First Edinburgh
  - Fife First
- First Glasgow (No1)
- First Glasgow (No2)
  - First Ayrshire
- First Hampshire
  - First Provincial Buses
  - First Southampton
- First Manchester
  - First Pioneer Bus
- First Midland Red Buses
- First Student (USA)
- First Transit (USA)
- First Wessex National
- First Western National Buses
  - North Devon
  - First Red Bus
- Leicester Citybus
  - First Leicester
- Mainline Group
  - First Mainline
- Midland Bluebird
  - First Edinburgh
- New World First Bus (26%)
- Northampton Transport
  - First Northampton
- PMT
  - First Crosville
  - First Pennine
  - First PMT
  - Red Rider
- Rider York
  - First York
- Walthamstow Citybus
  - First Capital
- Yorkshire Rider
  - First Bradford
  - First Calderline
  - First Huddersfield
  - First Leeds
  - First Superbus
  - First Quickstep

- First Great Eastern
- First Great Western
- First North Western
- Tramlink (Croydon) (20%)

New models at vibrant show

IT'S TEMPTING FOR organisers of regular exhibitions to bill each staging of the event as the best yet. But in the case of the Coach & Bus series of shows it seems that each event really does surpass its predecessor and Coach & Bus 99 was no exception.

There were new models from some of Britain's market leaders. The country's biggest coachbuilder, Plaxton, unveiled the Panther and Paragon coach range. And the biggest chassis maker, Dennis, previewed its new R-series coach chassis.

Other big names had new products too. Wright's new Eclipse and Solar stole the show as far as buses are concerned. Other new eye-catching single-deckers came from East Lancs, SC Coachbuilders and Mercedes-Benz. Volvo, whose double-deck prototype of two years ago has never actually turned a wheel in service, showed the new B7L, a model for which it claims over 450 orders. And the latest challenge to the supremacy of Dennis in the sub-12m single-deck bus market, the DAF SB120, had its first public showing.

In looking round the exhibition, held at the NEC, there was a strong FirstGroup presence. Vehicles in the group's corporate colours could be seen on no fewer than six stands - East Lancs, Marshall, Mercedes, Optare, Volvo and Wright. By contrast there was but one Stagecoach vehicle in the exhibition hall, an Alexander-bodied Dart, and similarly only one bus for an Arriva company in the shape of a London-specification dual-door Wright-bodied DAF SB120.

London operators were represented by Dennis Tridents for First Capital and Metroline, by two DAF SB220s for London Central's high-profile Millennium Dome operation, and the SB120 for Arriva. There were further London buses in the demonstration park, including Tridents for Stagecoach East London and Blue Triangle, and a Volvo B6BLE for London Traveller.

The new range from Wright, developed in co-operation with the bodybuilder's biggest customer, FirstGroup, marks a significant advance for bus design. The unusually-shaped deep windscreen is enough to make anyone take a second look at one in the street, while the stylish interior builds on the work done by FirstGroup and by other bodybuilders to give bus passengers a travelling environment more akin to a modern light rail vehicle. The articulated 56-seat (plus 92 standing) Eclipse Fusion was on Volvo's B7LA chassis, while the rigid 46-seat Solar was on a Scania L94. The artic weighed 17,951kg, making it the heaviest bus at the show - and suggesting that fuel consumption figures will prove interesting.

The structure too is new, using Aluminique aluminium extrusions which allow speedy replacement of side pillars after an accident, and improve structural strength by a claimed 25 per cent.

While the Wright bodies stole the show, there were other new models well worth a second look. East Lancs body designs have improved dramatically in the 1990s, shaking off the company's rather homely image. First there was the Cityzen (and one in the demonstration park was the only old-style step-entrance double-decker at Coach & Bus 99) and that was soon followed by other new models all with a wayward "y" in their names: Flyte, Spryte, Pyoneer, Lolyne and now Millennyum.

The Millennyum was unveiled at the show and proved to be a very attractive bus indeed - let down by a conservative interior, although that's an area dictated by London's needs rather than East Lancs' aspirations.

Indeed, when compared with the new Mercedes-Benz Citaro for FirstGroup, most observers reckoned that the Millennyum had the edge in overall styling appeal. The idea of East Lancs outdoing the mighty Mercedes is one which would have seemed laughable just a few short years ago.

Not that the Citaro was unattractive. But against the Solar, the Eclipse, and the Millennyum it appeared, let's say, a shade pedestrian. That it was only 2.5m wide, against the 2.55m of other new products, may have played a part in making it seem less spacious too. The Citaro replaces the O405N, and the bus on show was the first of 60 for FirstGroup.

On the Volvo stand there was a

The East Lancs Millennyum was one of the surprises of the show and is one of the most attractive bodies yet from the Blackburn coachbuilder. The show vehicles - there were two - were on DAF SB220 chassis for London Central.

B7L with Plaxton President body, one of 102 for Travel West Midlands. Alongside was what might be described as the Mark III version of the B6 - the B6BLE which Volvo will no doubt hope achieves more sales than its predecessors, the B6 and B6LE. The B6BLE was for FirstGroup, which has ordered 60, and differs from the previous B6LE in a number of respects - the most important being the adoption of independent front suspension.

Another variation on the B7L theme was a left-hand-drive 12.8m-long Volvo 7000 for operation in Helsinki. The 7000 is a low-floor citybus using a B7L underframe and bodywork built in-house by Volvo's Carrus subsidiary.

SC Coachbuilders (previously UVG, WS and originally Wadham Stringer) has inherited ... well, it has inherited not very much. Its predecessors failed to make much impact on the bus market, and the Compass on Dennis Dart SLF chassis, which SC took over from UVG, has found relatively few buyers - and although it wasn't on show it remains available. If it's possible for SC to make the grade as a bus builder, its new Nimbus is the product to do it.

Developed by parent company Caetano (and sold in Portugal as the Citadino), the Nimbus is a genuinely attractive alternative to the bodies being built on the ubiquitous Dart by the mainstream bodybuilders. It marks a move away from steel to aluminium for the structure, bringing SC Coachbuilders into line with other companies bodying the Dart.

Optare used Coach & Bus 99 to launch its revised Excel 2, which offers the option of a 230bhp Mercedes OM906 engine alongside the 160 and 215bhp versions of the Cummins B-series which powered the original. The Excel 2 has revised front and rear styling, the most noticeable change being twin round headlamps in place of the twin rectangular units of the original model. Excel 2 is offered in four lengths - 9.6m, 10m, 11m and 11.8m. The bus on show was a 40-seat 11.8m model in Trent colours. Optare also showed a Solo for First PMT.

Irisbus - the name for Iveco and Renault's merged bus operations - showed an Agora Line chassis. Built by Renault, the Agora Line will be bodied for UK operation by Optare using the standard Agora front and rear end structures. The first are to be completed in 2000. The Agora has a 9.8-litre Renault engine mounted in line with the chassis.

New on the DAF stand was the SB120 with Wright Cadet body - with one in the livery of Fishwick of Leyland while another in Arriva London red was being exhibited by Wright. The SB120 uses a Cummins/Allison drivetrain which will be familiar to anybody who has ever encountered a Dennis Dart. Imitation is said to be the sincerest form of flattery. DAF also showed a DB250 double-decker for A-Bus of Bristol, with 69-seat Alexander ALX400 body, and an SB220 with 43-seat Ikarus 481 body for the Solent Blue Line operation of Marchwood Motorways.

On the Plaxton stand the company's all-new coaches attracted most attention, but the company's best-selling Pointer was represented by two extremes - the 11.3m-long Super Pointer Dart, first shown at Coach & Bus 97, which weighed 7795kg and could carry up to 64 passengers (38 seated), and the newer 8.8m Mini Pointer Dart which entered production at the start of 1999 and has secured a remarkable number of sales. The MPD is a 47-passenger bus (26 seated) and weighed 6465kg. The Mini Pointer Dart is aimed at the same market as the Optare Solo.

Plaxton also showed a dual-door 69-seat President on a B7L to Lothian specification, but intended for long-term demonstration to Dublin Bus.

Aside from Plaxton's new models, the other main attractions for coach operators were the new Dennis R-series and the Bova Magiq. When Bova launched its bulbous-looking Futura back in 1983 it was quite radically different from anything else on the market. The sculpted front of the Magiq has a resemblance to the Futura but is more subtle. The high-set windscreen is out of step with current trends, but overall the Magiq maintains Bova's distinctive look. Power for the Magiq is DAF's recently-introduced XF 12.6-litre engine, rated at either 340 or 381bhp. A ZF 6S.1600 gearbox is standard. In its first year in the UK only five Magiqs are expected to be delivered, with a price tag of

The first right-hand-drive Mercedes-Benz Citaro integral, seen prior to being exhibited at the NEC. It is one of 60 for FirstGroup.

Wright's new bodies stole the show. This is the Solar on a Scania chassis. Note the partially-enclosed rear wheels.

around £190,000. The show Magiq was left-hand-drive. The Futura continues in production - with a minor facelift which adds round headlights.

The first 12m Caetano Enigma on a rear-engined chassis - a MAN 18.310 - was on display, as was a new 3.7m-high Berkhof Axial 70 body on a Volvo B10M for Moor Dale Coaches. Volvo showed an example of its B7R with Plaxton Prima bodywork, the 40th to enter UK service since the model was launched at Coach & Bus 97 - which doesn't really seem like a number to boast about.

At Coach & Bus 99 Plaxton exhibited a two-door President on a Volvo B7L chassis. Production of the attractive new model, at what was Northern Counties in Wigan, started at the beginning of the year.

Among small coaches, Euro Coachbuilders had its new Pantera model on the Mercedes-Benz Vario. The Pantera - Spanish for Panther - is presumably set for a race with the Plaxton Cheetah. Optare showed its new Soroco 16-seater, built in Spain on a Mercedes-Benz Sprinter chassis by Ferqui, who make the bigger Solera which is also sold by Optare. Ferqui, incidentally, is an amalgam of the names of the founders of the company - Fernandez and Quince.

Volvo did have a left-hand-drive B12T with Sunsundegui Sideral bodywork outside the hall, but was indicating that it had no immediate plans to import it to the UK. Ayats showed its spectacular Bravo double-decker, the first examples of which were imported to the UK in 1999. There were two on its stand, and a third in the demonstration park. Eye-catching it certainly is, but also something of a specialised taste.

Canon, based in County Tyrone, has been supplying small numbers of vehicles since it launched its Hi-line and Softline chassis in 1998. At Coach & Bus 99 it had a new 9.5m LowLine chassis in the demonstration park - yet another Dart competitor although hardly one to send shivers down the spines of the Dennis management team. If Dennis has been able to see off Volvo, and has yet to see any significant transfer of loyalty from the Dart to the new DAF SB120, it seems a bit unlikely that the Cannon LowLine will turn into a major threat.

Coach & Bus 99 illustrated the speed with which UK operators and manufacturers have switched to new low-floor bus designs, and it was the first major UK show not to feature an Olympian since the model's launch at the NEC in 1980. The air of optimism among exhibitors - supported by an impressive range of brand new products - augurs well for the future.

The first production B7Ls are being delivered to Travel West Midlands, which has 102 on order. Bodywork is by Plaxton.

The new DAF SB120 with Wright Cadet body, in Arriva's London livery. Wright is the only builder offering bodywork on DAF's copy of the Dennis Dart.

Bova's new Magiq was shown in left-hand-drive form. A small number of Magiqs will be sold in Britain in 2000, alongside the established Futura range.

New in Scotland

The big news in Scotland in 1999 was the introduction of Dennis double-deckers to the Lothian fleet, which had hitherto standardised on the Olympian and before that the Atlantean. These were Tridents, and the first five had Alexander bodies. The so-called harlequin livery marks a distinct break with Lothian's traditional madder and white.

Harte Buses runs one local service in the Greenock area, and does it with a fleet of immaculate modern buses. The newest is this Plaxton-bodied Dart.

Dart's Darts, was the inevitable comment when Dart Buses of Paisley bought its first Darts. They had Marshall Capital bodies. Manufacturer Dennis points out that as well as their being Dart Darts in Paisley, there are Dennis Dennises in Manchester .

Park of Hamilton added 10 Volvo B10Ms to its fleet at the end of the year for operation on Scottish Citylink services. They had Plaxton Premiere bodies.

Highland Country is part of the fast-growing Rapson group. New in 1999 was this Wright-bodied Volvo B10BLE.

Dennis launch rear-engined coach

For most coach operators the designation R-series conjures up a vision of a cheap and cheerful front-engined Ford. It's a model which was popular with many small (and some not so small) fleets for two decades, from its launch in 1965 to its demise in 1985.

Now there's a new R-series, which is as different from the old one as can be imagined.

The new R-series comes from Dennis and it's a sophisticated rear-engined underframe powered by a Cummins ISM 11-litre engine which is being offered in two power ratings - the 340bhp R340 and the 405bhp R405. The latter is the most powerful coach ever built in Britain, and it places Dennis firmly in the premier league.

The new model is not a conventional chassis but instead comprises front and rear modules, designed to be integrated into a completed coach structure. The lead bodybuilder is Plaxton - the project was being developed before Dennis was purchased by Mayflower in 1998 - and the new Paragon and Panther will be the first bodies to be offered on the R-series under an exclusive two-year agreement. When that expires Dennis will no doubt look to involve other coachbuilders.

The R-series has independent front suspension to give the best ride quality, and has disc brakes on all four wheels. The Cummins ISM engine is linked to an automated ZF AStronic 10-speed gearbox, with the option of the conventional six-speed ZF 6S1600 on the R340 or a six-speed Mercedes gearbox on the R405. Multiplex wiring is fitted to simplify the electrics.

Production of the R-series begins in 2000, although prototypes were on trial in 1999 and a complete coach with Plaxton Paragon body was used to launch the new model at Coach & Bus 99. It is an addition to the Dennis coach range, with the lower-powered mid-engined Javelin continuing in production. The R-series is designed for 3.5m-high bodywork, but can be bodied to a height of 3.2m although that will require a step up to the rear seats.

The story of Dennis in the UK bus market over the last decade has been one of continuing growth. The Javelin has usually been number two in coaches, after the Volvo B10M, although in 1999 it slipped to the number three position behind Volvo and Scania. Dennis will be hoping that the R-series will have the same success in the coach market that its Dart and Trident have had in the bus business.

website: www.dennisbus.com

The Dennis R-series will initially be available only with Plaxton bodywork. This is a Paragon.

1999 Roundup

January

• DA Travel of Stoke-on-Trent takes over the operations of Knotty Bus, best known in recent years for the operation of elderly AECs.

• The maximum gross vehicle weight for two-axled buses and coaches in the UK is increased from 17 to 18 tonnes.

• Martindale of Ferryhill closes.

• The unique 1992 Iveco TurboCity double-decker is advertised for sale in weekly trade magazine *Bus & Coach Buyer* at £39,950. It's been on sale since May 1998 - originally at £45,000.

• The first Frazer-Nash Eastleigh 100 is delivered to Anderson Travel of London. The Eastleigh 100 is a re-work of the Robin Hood Coachbuilders RH2000 body.

• National Express buys Robinson Bus Service Inc of Chicago. It operates 547 vehicles.

• Wright claim record output in 1998 of 565 buses and announce expansion plans to build up to 700 a year.

• Yorkshire Travel ceases operation in Huddersfield.

February

• Newmark Coaches and Guide Friday launch competing local services in Stratford-on-Avon as Midland Red South cuts back its operations.

• First double-deck Noge body, on an MAN 24.400 chassis, is delivered to Berryhurst of London.

• Stagecoach Manchester takes over Glossopdale, established in 1992 and running 21 buses.

• Lacey's of East Ham goes into liquidation. The old-established company ran 10 coaches.

• Carlton PSV Sales renamed Neoplan UK.

• Steele's Coaches of Annan goes into voluntary liquidation. It operated nine vehicles.

• Setra reports sales of a record 2,573 units worldwide in 1998, with German sales in excess of 1,200 vehicles.

• MTL North introduces four gas-powered Dennis Dart SLFs to park-and-ride services in Southport.

• The 1992 Iveco TurboCity double-decker - see January - now on offer at £35,000.

• Dunn-Line takes 80 per cent share in Executive Travel of Wednesfield (10 coaches) and buys Ascot of Derby (14 vehicles). Dunn-Line Derby is set up to run the former Ascot operations, founded in 1964.

• CMT Buses of Liverpool takes over ABC Travel of Formby (29 vehicles). ABC was set up in 1989 and runs a modern fleet which is made up primarily of Optares.

March

• Driver Express of Horsham goes into liquidation. It had been involved in the London Transport route 60 fiasco.

• Stagecoach Busways withdraws its last Atlantean, ending 39 years continuous operation of the type by Busways and its predecessors.

• Clayton Jones wins 50-vehicle O-licence for operations in London using a dormant company, Edwards of Pontypridd. The operations will be branded Eros Travel. Clayton Jones' wife, Alison, runs Shamrock Travel of Pontypridd.

• Classic Coaches of Annfield Plain buys the associated Moor-Dale and Primrose operations, running 23 vehicles between them. Moor-Dale had been owned by Proudmutual (parent of Northumbria Motor Services) until 1994. It acquired Primrose in 1997.

• MK Metro takes over Border Buses and Viscount Central of Burnley, which operate 59 vehicles between them. The Burnley companies had been in administrative receivership since October 1998.

• Fuggles of Benenden cease trading. The company ran 15 vehicles and had been in existence since the 1930s.

April

• Boon's of Boreham sells its coaching operations and five vehicles to Windmill Coaches of Colchester. Boon's retains buses for contract operations in Chelmsford.

• Duncan's Coaches of Huntingdon goes into liquidation. The business, originally family-run, had been owned since 1996 by Travel World Aviation and owned 15 coaches.

• Warrington Borough Transport closes its Coachlines operation.

• Rapson's Coaches of Inverness buys Orkney operator Shalders Coaches (40 vehicles).

• Renault takes a 22.5 per cent stake in Nissan Diesel, the bus and truck arm of the Japanese car maker.

• Euro Coach Builders of County Donegal unveils the Pantera body for Mercedes-Benz Vario chassis. It will replace the bigger EX33 on the Mercedes O1120L chassis, which is being discontinued.

• Western Isles Council starts operating its own 11-strong bus fleet on Lewis. It trades as Bus na Comhairle - Council Bus.

• Alf Moseley Continental launches the Continental 360, a 3.67m-high Marcopolo-bodied MAN 24.440. The first is for Yorks of Northampton, which is owned by Alf Moseley.

• Norfolk County Council decides to run its own buses on school tenders and orders 10 Dennis Javelins with SC Cutlass bodies.

• Green Triangle of Atherton takes over B&D of Leigh. B&D (Bellairs and Dootson) ran 14 buses.

• Wallace Arnold's 1999 new coach intake includes the company's 750th Volvo.

May

• Wilts & Dorset subsidiary Tourist Coaches acquires the business of Bell's Luxury Coaches, Winterslow (seven coaches). Bell's was established in 1921.

• East Kilbride operator White Ribbon closes, along with the associated Gilchrist Coaches business. White Ribbon had been running local services in East Kilbride since 1988.

• Coach of the Year at the 1999 Brighton coach rally is a Neoplan Starliner owned by Travelrich of Clacton. It is the first Starliner in Britain.

• Classic Coaches takes a 40 per cent share in Hylton Castle Coaches as part of a deal in which Hylton Castle's managing director acquires a 60 per cent interest in Classic's Moor-Dale subsidiary.

Coach of the Year at the 1999 Brighton coach rally was Britain's first Neoplan Starliner, owned by Travelrich of Clacton.

• ERF acquires the rights to build the ill-starred Marshall Minibus chassis.

• Plans for a £50million guided trolleybus route in Liverpool are turned down by a government inspector. The plan, developed by Merseytravel, would have seen articulated trolleybuses running from Albert Dock to Page Moss and to a park-and-ride site close to the M62 motorway. Merseytravel says the project won't be abandoned.

• National Express buys National Bus Company Pty, operating in Melbourne and Brisbane. It runs 794 vehicles and its subsidiaries include Westbus, which has a small British coach operation in Hounslow. It follows this up with the purchase of Transport Management Group Pty, operating in Perth with a fleet of 190 buses.

• New bus station opens in Sunderland.

• Burtons Coaches of Haverhill buys Premier Travel from Stagecoach.

• Tellings Golden Miller buys Capital Logistics (173 vehicles).

• C G Smith (Coachworks) of Long Whatton goes into receivership. The company built a small number midicoaches in the mid-1980s, including the Whippet on a front-engined MAN chassis.

June

• Alister's Coaches withdraws its services in Cardiff and Barry. The company had started running minibuses in Cardiff in 1998.

• The Midland Metro opens between Wolverhampton and Birmingham. The 12-mile line cost £145million and is almost 12 months late in opening.

• National Express buys Silverwing of Heathrow (88 vehicles). It is put under the control of Airlinks (formerly Speedlink).

• National Express wins franchises to operate train and tram services in Australia - V/Line Passenger, Bayside Trains and Swanston Trams, all in Victoria.

• Mayflower announces a joint venture in the USA with Freightliner, part of the DaimlerChrysler group. It will see Freightliner subsidiary Thomas Built Buses producing Alexander-bodied Dennis Darts at a new factory in North Carolina. The joint venture company is Thomas Dennis Co. A British built bus, branded as the Thomas SLF200, is exhibited at the American Public Transit Association exhibition in October.

• New operator Mac Tours launches Edinburgh sightseeing tours using classic open-top Titans and Lodekkas.

• MAN is to join the competitive UK midibus battle with its 14.220 chassis.

• Davies of Pencader (65 vehicles) goes into receivership. The bus operations and 38 vehicles are taken over by First Cymru.

• Cave's Bus Services of Solihull closes. It operated 12 buses. The company had been started in 1948. Its closure follows the death of the owner.

• DA Travel of Stoke-on-Trent ceases trading. Its fleet had numbered 24 vehicles at the beginning of the year.

The distinctive double-deck integral Ayats Bravo I marks the Spanish builder's return to the UK after 20 years.

• The first Ayats Bravo is delivered to a UK operator.

• Seven members of the Van Hool family resign from the company's board, prompting speculation - hotly denied - that the company is ripe for a take-over bid. Van Hool was established in 1947, and built 1,700 buses and coaches in 1998.

• Mercedes-Benz announce new Travego coach to replace the O404 integral, launched in 1992. Target output is 600 a year from 2000. There will be no right-hand-drive version. Operators in Britain and other right-hand-drive territories will still be able to buy O404 underframes - in the UK's case with Hispano Vita bodywork.

• Status Bus and Coach Group is formed, creating a coalition of operators in which the original owners retain a majority shareholding in their companies, as well as being directors and shareholders of Status. The Status group's members run 600 coaches and buses between them. (For a listing, see p43)

July

• Yorkbus is set up by a former Lothian Region Transport director, running six ex-Lothian Atlanteans on local services in York. It closes after four weeks.

• SC Coachcraft of Rotherham, maker of the Vision midicoach on Mercedes chassis, goes into liquidation. It is not associated with similar-sounding SC Coachbuilders of Waterlooville, owned by Caetano.

• Timeline Travel of Bolton acquires Barry Raybould of Walsall (nine coaches)

• Yorkshire Traction takes a 25 per cent share in London Traveller - setting up a new company, Metropolitan Omnibus (London) which will continue trading under the London Traveller name. The original company was set up in 1996. Its operations outside London are not included in the deal and are to be sold separately.

• Dennis delivers its 3,000th low-floor Dart SLF. It is one of 13 for Armchair of London.

August

• National Express buys Texas-based Durham Transportation (3,500 vehicles) and Kenneth E Bauman Bus of Pennsylvania. It now runs 4,000 school buses in the USA.

• Irisbus announces that it is taking a 57.7 per cent share in Hungarian bus builder Ikarus, which builds around 1,100 vehicles a year.

• Dunn-Line buys Camms of Nottingham.

• Three buses are destroyed by a fire at First Leicester's depot.

September

• Henlys buys US manufacturer Blue Bird Corporation, which builds 14,000 buses a year. Blue Bird was founded in 1932 and is the leading builder of school buses in North America. Henlys plans to increase Blue Bird's export sales in conjunction with Volvo, which has sales outlets in 60 countries,

• The Rapson Group buys two Orkney businesses - Rosie of Evie and Peace of Kirkwall. Rosie runs three coaches; Peace has 30.

• The Go-Ahead Group buys Metrobus of Orpington for £14.6million (200 buses).

• Coachmasters of Rochdale wins school contracts in Greater Manchester and launches a Rochdale to Wigan bus service.

• Woods of Falkirk gives up its local bus operations. First Edinburgh takes over.

• Optare announces that it is to build a new 7.2m-long low-floor 16-seater aimed at specialist welfare and rural bus operators. An artist's impression of the un-named minibus with a composite moulded one-piece body shows a vehicle with styling more akin to a people-carrier than a bus. Production is to start in 2000.

• Henlys establishes a small design and engineering centre in Hungary, prompting speculation that it may use Hungary to create a low-cost manufacturing base which would increase the group's production capacity.

• A coach crash in South Africa kills 26 British tourists. The vehicle was owned by Springbok Atlas Tours.

October

• The last Volvo Olympian chassis leaves the production line at Irvine - it is one of seven for Blazefield Holdings. Volvo claim total Olympian production of 10,257, including those built by Leyland.

• Anglo American Coach Industries, importers of the Ayats coach range to the UK, announce plans to import from Malaysia a 12m citybus, the aiRide, and two new coaches - the 3.5m-high Montana and 3.7m-high Colerado. The coaches will be built on MAN 18.310 underframes, and five are reported to have been ordered by UK fleets.

• Skills of Nittingham takes over local operator Dolphin Coacjes as well as Crusader of Barnsley. No coaches are acquired in either deal.

• Translinc of Lincoln buys Elsey of Gosberton (12 vehicles). Translinc is the former Lincolnshire county council transport operation, privatised in a management-led employee buy-out in 1998.

• Wright of Ballymena teams up with US manufacturer Chance Coach to sell buses in North America. Wright will provide a version of the new Solar/Eclipse bodies unveiled at Coach & Bus 99, while Chance will build a new low-frame chassis.

• Partridge of Hadleigh closes. Shortly before the planned closure four vehicles are destroyed by a fire at the company's depot.

• Airlinks, part of the National Express group, buys Cambridge Coach Services from Blazefield. CCS operates 22 vehicles.

• Berkhof launches a double-deck coach body, the Axial 100DD. The first UK order is placed by Siesta International of Middlesbrough for delivery in 2000.

• The East Lancashire Partnership publishes plans for a rapid transit system based on the two main rail routes in the region - Preston to Colne and Clitheroe to Manchester - with new high-quality connecting bus services at interchanges in Accrington, Blackburn and Burnley. The report envisages buses powered by LPG - or even trolleybuses for some routes.

• Williams of Deiniolen sells its bus services to Arriva Cymru but continues as a contract operator. One bus - a Mercedes mini - passes to Arriva.

• The unique Iveco TurboCity double-decker (see February) finally finds a home with Bluebird of Manchester.

November

• The Kings Ferry buys the coach operations of Stagecoach East Kent.

• Dunn-Line of Nottingham buys nine-vehicle APA Travel of London - perhaps best known for its Spanish Speaking Services fleet - and the 20-vehicle Stevensons coach operation of Arriva Midlands North. Dunn-Line Holdings now runs almost 150 coaches.

• New operator Travelspeed, trading as Northern Blue, starts operating between Accrington and Darwen using elderly Leylands - Atlanteans, Nationals and Tigers.

• Robson of Thornaby gives up local bus operation. The company operated in Stockton, primarily using Leyland Nationals.

• Two Lanarkshire coach operators - Bruce of Shotts and Long of Salsburgh - launch M8 Motorvator express between Glasgow and Edinburgh.

• Registrations for local services in Scarborough operated by Applebys are cancelled by the traffic commissioner because of continued vehicle maintenance problems. A 12-month ban is imposed, and the company's operating licence reduced from 90 to 60 vehicles.

• The 10,000th Olympian enters service - with Dublin Bus.

• 35 years of Atlantean operation by Nottingham City Transport comes to an end.

• New operator Glasgow Citybus starts up, running second-hand Dennis Darts.

• Lothian Region Transport buys the Guide Friday service between Edinburgh city centre and the airport, which ran in competition with LRT. This follows earlier deals between the two operators which saw LRT give up its tour operations in Cambridge and York.

December

• Dunn-Line of Nottingham buys Seamarks of Luton (20 vehicles).

• Rapsons of Inverness takes over Dunnett of Keiss (12 vehicles). The Rapson group now runs 250 buses and coaches.

• Town & Country Travel takes over the 30-vehicle Essex local bus operations of Ensign Bus.

• Wakes of Wincanton (32 vehicles) is taken over by Huberts of Yeovil.

• Clayton's Coaches of Leicester, formed in 1920, closes after the death of its owner.

• Harris Bus goes into administrative receivership after a rumoured sale to FirstGroup falls through. The company operates 76 buses on LT contracts. Frank Harris Coaches is not affected.

• National Express buys the ATC Group, claimed to be the USA's largest privately-owned bus operator running 2,000 vehicles. It boosts National Express's US fleet to 8,000 vehicles.

• Ford launches new Transit.

• US builder Metrotrans, part owned by Mayflower, files for bankruptcy protection.

End of an era at Northern Counties

NORTHERN COUNTIES - the name, not the factory - disappeared in 1999. Out went the Palatine, in came the President. And the President is a Plaxton product.

Yet in 1999 Plaxton Wigan - the new name for Northern Counties - built over 350 double-deckers, which is quite an achievement when you consider that new bus sales are lower now than they were, say, half a century ago. But half a century ago Northern Counties was dwarfed by such giants as Metro-Cammell and Eastern Coach Works. Today it vies with Alexander to be the top double-deck producer.

Northern Counties started building bus bodies in Wigan in 1921, its first double-decker coming in 1928 - unusually, on a Maudslay chassis. It found customers among a wide range of operators and was one of the builders of utility buses during World War II. In the postwar years it established itself as a supplier to a number of BET companies, perhaps most famously Southdown with its distinctive PD3 Titans.

It was the major supplier to local independent Lancashire United Transport, and to a number of small and medium-sized municipals, generally concentrating on double-deck bodies.

In 1967 it took over the rival Massey Bros business in Wigan, and it is in fact the former Massey site in Pemberton which is the basis of the present modern Plaxton Wigan factory.

Northern Counties worked closely with the newly-created SELNEC PTE in 1970 in the development of the SELNEC standard double-decker, and soon became the major supplier to SELNEC and its successor, Greater Manchester. This relationship developed to the point where the PTE bought the company in 1985.

Concerns about the supply of chassis from Leyland, which was developing an integral double-decker, led to the appearance of the Foden-NC in 1975, a project which briefly saw Foden re-appear as a builder of double-deck chassis. The project did not proceed, nor did later plans to have Northern Counties build Leyland's integral Titan under licence.

The troubled times of the mid-1980s took their toll, even though Northern Counties developed a minibus body which it built in respectable quantities from 1986. There was a brief tie-up with Renault under which the company built French-style bodies on the PR100 in 1988. That the PR100 failed was no reflection on Northern Counties - it just wasn't the right product for UK operators. Only five were built. The company was ready to develop a rear-entrance version of its double-deck body for a projected new-generation London Transport Routemaster in 1991, but that never saw the light of day.

As orders slowed down the administrative receivers were called in during 1991, following which the company was the subject of a management buy-out in 1992.

The Countybus name was used briefly in 1991 for a new single-deck body, but in 1992 this was rechristened the Paladin. The standard double-deck was briefly called the Countydec before becoming the Palatine. The Pageant name was applied to the minibus - but production of this ceased in 1992 as orders for big buses picked up.

The Palatine II with its deep curved windscreens appeared in 1993. Henlys, owners of Plaxton, took over in 1995.

In the 1990s Northern Counties was supplying Palatines to all of the big groups - Arriva, FirstGroup, Go-Ahead, MTL and Stagecoach - and to a number of other operators too including Capital Citybus, East Yorkshire and Metrobus.

For the low-floor market Northern Counties developed a new Paladin, which was unveiled at Coach & Bus 95. Production of this was transferred to Plaxton at Scarborough in 1997 where it became the Plaxton Prestige. What are likely to be the last Prestiges, on DAF chassis for Arriva, were built in Scarborough in 1999.

Under Henlys ownership the factory has been transformed, with major investment in new production facilities, now dedicated to the manufacture of the President double-decker.

During 1999 Plaxton Wigan delivered Presidents to Arriva London, Capital Logistics, Dublin Bus, Lothian, Metroline and Travel West Midlands. To this list in 2000 it will be adding London General and London United.

website: www.plaxton.co.uk

Although little-used, the Countybus name was still being applied to the last of the Palatines.

All change at Wigan. On the production line at the start of the year were the last Northern Counties Palatine bodies (left) and some of the first Plaxton Presidents.

Who owns whom...

A quick guide to the major groups and their key subsidiary companies, associated companies and principal trading names.

Appleby
- R W Appleby
- Fleetjet
  - North Bank Travel
- Halcyon Leisure

Arriva - see page 19

Blackpool Transport Services
- Handy Bus

Blazefield Holdings
- Harrogate & District Travel
- Huntingdon & District
- Keighley & District Travel
- Premier Buses
- Sovereign Bus & Coach Co
- Sovereign Buses (London)
- Yorkshire Coastliner

Bournemouth Transport
- Christchurch Buses
- Dorset Travel Services
- Vintage Yellow Buses
- Yellow Buses
- Yellow Coaches

CMT Buses
- ABC Travel

Durham Travel Services
- York Pullman

EYMS Group
- East Yorkshire Motor Services
- Finglands
- Scarborough & District Motor Services

FirstGroup - see page 29

Flights Travel Group
- Excelsior Coachways
- Flights Coach Travel

Go-Ahead Group
- Brighton & Hove Bus and Coach Co
- City of Oxford Motor Services
  - The Oxford Bus Company
  - The Wycombe Bus Company
- Go North East
  - Go Gateshead
  - Go Xpress
  - Go Northern
  - Go Wear Buses
  - Go Coastline
- London Central Bus Co
- London General Transport Services
- Metrobus
- *Thameslink (65%)*
- *Victory Railways (65%)*
  - *Thames Trains*

London United Busways
- Airbus
- Stanwell Buses
  - Westlink

Metroline Holdings
- Brents Travel
- Metroline London Northern
- Metroline Travel
- Scottish Citylink
- Skye-ways (25%)
- West Coast Motors (25%)
  - Oban & District

Moseley Group
- Bowens
- Yorks

MTL Services
- Merseyside Transport
- MTL Heysham
  - Heysham Travel
- MTL North
- Southport & District
- Wirral Peninsula
- *Merseyrail Electrics*
- *Northern Spirit*

National Express Group
- Airlinks The Airport Coach Company
  - Cambridge Coach Services
  - Jetlink
  - Silverwing
  - Speedlink
- ATC Group (USA)
- Kenneth E Bauman Bus (USA)
- Crabtree-Harmon (USA)
- Durham Transportation (USA)
- Eurolines (Nederland)
- Eurolines (UK)
- Flightlink
- Group Bronckaers (Belgium)
- National Bus Company (Australia)
  - Westbus (UK)
- National Express
  - Rapide
- National Expressliners
- Robinsons Bus Service Inc (USA)
- Taybus Holdings
  - Tayside Public Transport Co
  - Travel Dundee
  - Travel Greyhound
  - G&M Wishart
- Transport Management Group (Australia)
- West Midlands Travel Group
  - Smiths Coaches (Shennington)
  - Travel London
  - Travel Merryhill
  - Travel Your Bus
  - Travel West Midlands
  - West Midlands Travel
- *Central Trains*
- *Gatwick Express*
- *London & Continental Railways (17%)*
- *Midland Mainline*
- *ScotRail*
- *Silverlink*
- *Travel Midland Metro*

Northern Ireland Transport Holding Company/Translink
- Citybus
- Flexibus
- Ulsterbus
  - BusyBus
  - Goldliner
- *NI Railways*

Nottingham City Transport
- Pathfinder
- South Notts

Parks of Hamilton
- Trathens Travel Services

Pete's Travel
- Busy Bus Co
- Lionspeed

Bournemouth Transport Yellow Buses added nine East Lancs Lolyne bodied to its fleet during 1999. ***John Marsh***

Wilts and Dorset is one of the smaller companies which purchase vehicle types not chosen by the large groups. Three low-floor DAF Optare Spectras were delivered in 1999. *John Marsh*

Rapson group
- Highland Bus & Coach
- Highland Country Buses
- Peace's Coaches
- Rapson's Coaches
- Shalders Coaches

Reading Transport
- Goldline Travel
- Greater Reading Omnibus Co
  - Reading Mainline
- London Line
- Newbury Buses
- Reading Buses

Rossendale Transport
- Ellen Smith Coaches

Southern Vectis
- Musterphantom
  - Solent Blue Line
- Southern Vectis Omnibus Co

Stagecoach - see page 7

Status Bus & Coach
- Bakers Coaches, Biddulph
- Border Buses, Burnley
  - Viscount Central Coaches
- Burtons Coaches, Haverhill
  - Premier Travel
- Capital Logistics, London
- Classic Coaches, Annfield Plain
  - Primrose
- Hylton Castle, South Hylton
- Link Line, London
- The Londoners, London
- MK Metro, Milton Keynes
- Midland Choice Travel
- Moor Dale Coaches, South Hylton
- Tellings Golden Miller, London

Traction Group
- Andrews (Sheffield)
  - Yorkshire Terrier
- Barnsley & District Traction Co
- Lincolnshire Road Car Co
- Meffan (Kirriemuir)
- Metropolitan Omnibus (London) (25%)
  - London Traveller
- Strathtay Scottish
- Yorkshire Traction Co
  - Coachlink
  - Fastlink
  - Townlink

Wellglade
- Barton Buses
- Kinchbus
- Nottinghamshire & Derbyshire
  - Blue Apple
- Trent Buses

Wilts & Dorset
- Hants & Dorset Motor Services
  - Damory Coaches
- Levers Coaches
- Tourist Coaches
  - Bell
- Wilts & Dorset Bus Co

EvoBus
- Mercedes-Benz
- Setra

Henlys plc
- Blue Bird (USA)
- Kirkby Coach & Bus
- Plaxton Scarborough
- Plaxton Wigan
- Prevost (Canada) (49%)
  - NovaBus
- Roadlease
  - National Expressliners

Irisbus
- Ikarus (57.7%)
- Iveco
- Heuliez
- Karosa
- Orlandi
- Renault

Mayflower
- Alexander
- Dennis
- Metrotrans (USA) (40%)
- Thomas Dennis (49%)

Optare
- Autobus Classique

VDL Group
- DAF Bus
- Berkhof
- Jonckheere
- Smit

Volvo Bus
- Carrus Oy
- Drogmoller
- Henlys (10%)
- MASA
- Prevost (Canada) (51%)
  - NovaBus
- Steyr
- Volvo Coach Sales

Mayflower, having won the battle with Henlys to buy Dennis in 1998 is able to supply complete vehicles such as this Alexander bodied Dart for Stagecoach Cheltenham and Gloucester. *John Marsh*

New in the Midlands and Wales

New - but not for long - in the Midlands was the Stratford local bus operation of Newmark Coaches, for which the company added Marshall-bodied Dennis Darts to its fleet at the start of the year. Launched in February, the services ceased in September.

Paul S Winson runs one of Britain's best-kept coach fleets, and runs a small number of buses too. A new Mercedes joined the fleet in 1999, with Plaxton Beaver 2 body.

The combination of Plaxton Premiere body and DAF SB3000 chassis is relatively uncommon. Most 1999 Premieres were on Volvo or Dennis chassis. Silverdale of Nottingham was one buyer. Note that it carries a number in the same T-AUA block as the Ludlows DAF opposite, an indication that both were registered by the supplying dealer.

Major investment by Phil Anslow Travel saw the addition of 13 new Mercedes-Benz Varios to the company's fleet. They had Plaxton Beaver 2 bodies.

Glyn Williams operates a network of bus services in South Wales and new vehicles in 1999 included four Mini Pointer Darts in a silver livery which incorporated a Welsh dragon to mark the company's 25th anniversary.

Ludlows of Halesowen was one of many small operators around the country to take advantage of the availability of Ikarus-bodied DAFs from Arriva Bus & Coach. It took one, as well as buying a Wright-bodied Dart.

Irisbus - a new name in buses

On 1 January 1999 Renault and Iveco combined their bus manufacturing operations in a new organisation initially known as IRIS.BUS. As the year progressed this metamorphosed first into Iris.Bus and finally to Irisbus. And it isn't pronounced like the flower - correctly it should be pronounced in the Italian way as "Eerisbus".

It adopted a stylised dolphin as its logo which, to quote the company, expresses its values - "Swiftness, humanity, intelligence, respect for people and the environment". What more can be said?

The two partners are of course well-known in the UK. Renault is best remembered for its minibuses - the S46, S56 and S75 - which were manufactured at Dunstable and were originally branded as Dodges. The last entered service in 1993 and small numbers are still around. Renault did also try to sell its full-size PR100 in 1988, but with no great success. Two can be seen running for Hornsby of Scunthorpe.

Iveco, too, came to prominence with minibuses. Where the Renault mini was in the main bought by PTE and municipal fleets, Iveco's minis went to NBC companies and to independents. Iveco still sells minibuses, but in small numbers and primarily to the welfare sector.

Iveco has made vain attempts to sell big buses in the UK - first with the TurboCity and then with the EuroRider. But it has succeeded in the coach market, with the combination of EuroRider chassis and Spanish-built Beulas body. It also sells small numbers of the quaint EuroMidi coach, bodied in Spain by Indcar.

Irisbus will continue to sell Iveco coaches in the UK, and announced plans to break into the bus market with the Renault Agora Line, to be bodied by Optare. The rear-engined Agora Line chassis was unveiled at Coach & Bus 99. It has a 9.84-litre Renault engine mounted in-line at the rear and inclined to an angle of 68 degrees. It is offset to the left side of the chassis, as with Volvo's B7L.

On the Agora Line Irisbus will be offering two power ratings - 206 or 256bhp - and a choice of four- or five-speed ZF automatic gearboxes. Irisbus anticipates that an Optare-bodied single-door Agora Line will have 48 seats with space for one wheelchair.

Irisbus has a distinguished heritage. On the Iveco side it embraces once well-known bus builders such as Fiat, OM, Pegaso and Magirus-Deutz, while over the years Renault has absorbed Berliet, Chausson, Saviem and, most recently - in 1998 - Heuliez. It has production facilities in France, Italy and the Czech Republic - to which in July 1999 it added a 57.7 per cent majority interest in Hungarian builder Ikarus.

In 1998 the Irisbus companies built 6,062 vehicles over 6 tonnes (3,346 from Renault and 2,716 from Iveco) and were the major suppliers of buses and coaches in France and Italy, and a significant force in Spain.

Add to this Ikarus production and vehicles under 6 tonnes, and Irisbus claims annual output of 3,700 buses, 2,600 coaches, 1,400 chassis and 1,800 minibuses - a total of 9,500 vehicles.

One of the more unusual products which is now part of the Irisbus stable is the Iveco EuroMidi, sold in Britain with Indcar Maxim bodywork. This 29-seater was delivered to Jeffs Coaches of Brackley.

Deliveries 1999

Significant new vehicle deliveries during 1999 included:

A1A, Birkenhead	2	Optare Solo	
A-Bus, Bristol	1	DAF DB250	Alexander ALX400
ABC, Formby	2	Optare Solo	
Airlinks, Heathrow	6	Dennis Dart SLF	Plaxton Pointer
	26	Volvo B6BLE	Wright Crusader
Alexcars, Cirencester	1	Dennis Dart SLF	SC Compass
Ambassador, Norwich	7	Volvo B10BLE	Wright Renown
Andrews, Sheffield	15	Dennis Dart SLF	East Lancs Spryte
Anglian, Loddon	4	Dennis Dart SLF	Plaxton Pointer 2
APCOA, Heathrow	12	Scania L94	Wright
Applebys, Louth	1	Scania L94UB	Wright
Armchair, London	13	Dennis Dart SLF	Alexander ALX200
Arriva	17	DAF SB220 LPG	Plaxton Prestige
	107	Dennis Dart SLF	Plaxton MPD (b)
	41	Volvo B6BLE	Wright (b)
Arriva London	20	DAF DB250	Plaxton President
Arriva North East	6	DAF SB220	Ikarus 481
Arriva Yorkshire	18	DAF DB250	Optare Spectra
BC Transit, Canada	51	Dennis Dart SLF	Plaxton Pointer 2
Bennetts, Gloucester	1	DAF SB220	Ikarus 481
Bharat, Birmingham	3	Dennis Dart SLF	Plaxton Pointer 2
Blackburn Transport	5	Volvo B10BLE	Wright Renown
	10	Dennis Dart SLF	Plaxton MPD
Blackpool Transport	9	Optare Excel	
Blue Bus, Horwich	2	Volvo Olympian	East Lancs Pyoneer
	2	Dennis Dart SLF	Plaxton Pointer 2
Blue Triangle	9	Dennis Trident	East Lancs Lolyne
Bluebird, Manchester	1	Scania N113	East Lancs Cityzen
	3	Dennis Dart SLF	Wright Crusader
Boulton, Church Stretton	1	Optare MetroRider	
	2	Optare Solo	
Brighton & Hove	20	Dennis Trident	East Lancs Lolyne
British Airways, Heathrow	2	DAF SB220	Ikarus 481
Bullock, Cheadle	4	Volvo Olympian	NC Palatine II
Bus Eireann	10	Volvo B10L	Wright Liberator
	10	Dennis Dart SLF	Plaxton SPD
	33	Mercedes O404	Hispano Vita
	23	Scania L94	Irizar Century
	12	Volvo B10M	Plaxton Excalibur
Canavan, Kilsyth	2	Dennis Dart SLF	Plaxton MPD
Capital Logistics	10	DAF DB250	Plaxton President
	6	DAF DB250	Optare Spectra
Cardiff Bus	20	Dennis Dart SLF	Plaxton SPD (a)
Cawlett group	2	Dennis Dart SLF	East Lancs Spryte
	4	Dennis Dart SLF	Plaxton Pointer 2
	8	Mercedes Vario	Plaxton Beaver 2
Citybus, Hong Kong	30	MAN dd	Berkhof-Jonckheere
Cityspeed, Birmingham	1	Dennis Dart SLF	Plaxton Pointer 2
Claribel, Birmingham	9	Optare Excel	
	1	Optare MetroRider	
Dart Buses, Paisley	3	Dennis Dart SLF	Marshall Capital
Davidson, Bathgate	2	Dennis Dart SLF	Plaxton Pointer 2
Delaine, Bourne	1	Volvo Olympian	East Lancs Pyoneer
De Luxe, Mancetter	2	Dennis Dart SLF	Plaxton Pointer 2
Dennis, Ashton	2	Dennis Dart SLF	Plaxton Pointer 2
Dublin Bus	195	Volvo Olympian	Alexander R-type
	2	Dennis Trident	Alexander ALX400
	1	Volvo B7L	Plaxton President
	20	Volvo B6BLE	Wright (b)
Dunn-Line, Nottingham	1	Scania N113	East Lancs Cityzen
Eastbourne Buses	1	Dennis Dart SLF	SC Compass
	1	DAF SB220	Ikarus 481
Epsom Buses	13	Dennis Dart SLF	Plaxton MPD
Express Travel, Liverpool	1	Dennis Dart SLF	East Lancs Spryte
Finglands, Manchester	2	Mercedes O405N	Mercedes
First Beeline	7	Volvo B12T	Plaxton Excalibur
First Capital	54	Dennis Trident	Plaxton President
First CentreWest	20	Volvo Olympian	NC Palatine I
	31	Dennis Trident	Plaxton President
	74	Dennis Dart SLF	Marshall Capital
First Leeds	15	Volvo B10LA	Wright Fusion
First Mainline	70	Volvo B10BLE	Wright Renown
First Manchester	15	Volvo B10LA	Wright Fusion
First PMT	16	Optare Solo	
First Southern National	6	Dennis Dart SLF	East Lancs Spryte
Flyerbus, Dublin Airport	2	DAF SB220	Ikarus 481
Frodingham Coaches	2	Dennis Dart SLF	SC Compass
Go-Ahead London	44	Dennis Dart SLF	Marshall
	11	Dennis Dart SLF	Plaxton Pointer 2
Go-Ahead North East	10	Optare Excel	
	14	Optare Solo	
Grayline, Dublin	1	Volvo Olympian	East Lancs Pyoneer
Halton Transport	4	Dennis Dart SLF	Marshall Capital
Hedingham Omnibuses	1	Volvo Olympian	Alexander R
Highland Country	1	Volvo B10BLE	Wright Renown
	1	Volvo B6BLE	Wright Crusader
	1	Dennis Dart SLF	Plaxton Pointer 2
	1	Optare MetroRider	
Hutchison, Overtown	3	Optare Excel	
Ipswich Buses	3	Dennis Dart SLF	East Lancs Spryte
Kavanagh, Urlingford	2	Volvo Olympian	NC Palatine I
Keighley & District	12	Volvo B10BLE	Wright Renown
Kettlewell, Retford	1	Scania N113	East Lancs Cityzen
KMB, Hong Kong	60	Volvo Olympian	Alexander ALX500
KMP, Llanberis	1	Dennis Dart SLF	Plaxton Pointer 2
Leask, Lerwick	4	DAF SB220	Ikarus 481
Limebourne	34	Dennis Dart SLF	SC Compass
London Traveller	15	Volvo B6BLE	East Lancs Spryte
London United	104	Dennis Dart SLF	Plaxton Pointer 2
Lothian	25	Dennis Dart SLF	Plaxton SPD (b)
	40	Dennis Trident	Plaxton President
	5	Dennis Trident	Alexander ALX400
	1	Dennis Javelin	Berkhof
Ludlows, Halesowen	1	DAF SB220	Ikarus 481
	1	Dennis Dart SLF	Wright
Marchwood Motorways	2	DAF SB220	Ikarus 481
Mayne, Manchester	5	Dennis Trident	East Lancs Lolyne
Metrobus, Orpington	15	Dennis Trident	East Lancs Lolyne
	4	Dennis Dart SLF	Plaxton MPD
	3	Dennis Dart SLF	Alexander ALX200
Metroline	65	Dennis Trident	Plaxton President
	50	Dennis Trident	Alexander ALX400
Midland Choice	1	Optare MetroRider	
Mitcham Belle	12	Dennis Dart SLF	Plaxton Pointer 2
MK Metro	8	Optare Solo	
MTL North	75	Dennis Dart SLF	Marshall Capital
	4	Dennis Dart CNG	Plaxton SPD
Nefyn Coaches	1	Optare Solo	
Newmark, Stratford	4	Dennis Dart SLF	Marshall Capital
Newport Transport	6	Dennis Trident	Alexander ALX400
Nip-On, St Helens	1	Dennis Dart SLF	Marshall Capital
Norfolk County Services	10	Dennis Javelin	SC Cutlass
Norfolk Green	1	Mercedes O405N	Mercedes
Nottingham City Transport	4	Volvo Olympian	East Lancs Pyoneer
Nova Scotia, Winsford	4	Dennis Dart SLF	Plaxton MPD
	1	Dennis Dart SLF	Marshall Capital
Oxford Bus Company	20	Dennis Trident	Alexander ALX400
	15	Volvo B10BLE	Wright Renown
	5	Volvo B10M	Plaxton Excalibur
Parks, Hamilton	20	Volvo B10M	Jonckheere Mistral
	18	Volvo B10M	Plaxton Premiere
	10	Volvo B10M	Van Hool Alizee
Patterson, Birmingham	2	Dennis Dart SLF	Wright Crusader
Petes, Smethwick	11	Dennis Dart SLF	Plaxton Pointer 2
Pink Elephant, Heathrow	10	Optare Solo	

Powells, Rotherham	2	Dennis Dart SLF	Plaxton MPD
Plymouth Citybus	13	Dennis Dart SLF	Plaxton SPD
Reading Buses	9	Optare Excel	
	7	Optare Solo	
Red Rose, Aylesbury	1	Optare MetroRider	
Road Car	5	Dennis Dart SLF	Plaxton MPD
	6	Dennis Dart SLF	East Lancs Spryte
	3	Scania L94	Wright
Scarlet Band	3	Dennis Dart SLF	Plaxton MPD
Shalder	2	MAN 11.220	Marshall
Shamrock, Pontypridd	3	Scania N113	East Lancs Cityzen
	1	Dennis Dart SLF	Marshall Capital
Shearings	16	Volvo B10M	Van Hool
	16	Volvo B10M	Jonckheere
Shuttle Buses, Kilwinning	1	Optare MetroRider	
Silver Star	1	Dennis Dart SLF	Plaxton Pointer 2
Simonds, Botesdale	2	Dennis Dart SLF	Plaxton Pointer 2
Solent Blue Line	8	Dennis Trident	East Lancs Lolyne
South Lancs Transport	3	Dennis Dart SLF	Plaxton MPD
Sovereign (London)	57	Dennis Dart SLF	Plaxton Pointer 2
Springfield, Wigan	3	Dennis Dart SLF	Plaxton MPD
Stagecoach group	6	Volvo B10M artic	Jonckheere Modulo
Stagecoach Oxford	27	MAN 24.350	Jonckheere Monaco
Stagecoach Portugal	10	Dennis Dart SLF	Camo
Stockton Borough Council	2	Optare Excel	
Stones, Leigh	4	Dennis Dart SLF	Plaxton MPD
Strathtay	4	Dennis Dart SLF	East Lancs Spryte
Swanbrook, Cheltenham	1	Dennis Dart SLF	Marshall Capital
Thamesdown	10	Dennis Dart SLF	Plaxton SPD
Tillingbourne	2	Dennis Dart SLF	Plaxton Pointer 2
Translink, Ulster	90	Volvo B10BLE	Wright Renown (b)
	5	Dennis Dart SLF	Wright Crusader
Travel West Midlands	80	Mercedes O405N	Mercedes
	11	Mercedes 0405GN	Mercedes
	45	Optare Solo	
	35	Optare Excel	
	20	DAF DB250	Optare Spectra (a)
	102	Volvo B7L	Plaxton President (b)
Trent	25	Optare Excel	
	4	Dennis Dart SLF	Plaxton Pointer 2
Truronian	4	Dennis Dart SLF	Plaxton MPD
UK North, Manchester	4	DAF DB250	Alexander ALX400
University Bus	1	Dennis Dart SLF	Wright Crusader
Warrington	8	Dennis Dart SLF	Marshall Capital
	2	Optare MetroRider	
Whippet, Fenstanton	1	Volvo B10BLE	Alexander ALX300
Wilfreda Beehive	2	Scania N113	East Lancs Cityzen
Williams	4	Dennis Dart SLF	Plaxton MPD
Wilts & Dorset	3	DAF DB250	Optare Spectra
	18	Optare Solo	
	3	DAF SB3000	Plaxton Premiere
Wolverhampton University	5	Volvo B6BLE	Wright Crusader
Worths, Enstone	1	Dennis Dart SLF	Plaxton Pointer 2
Yellow Buses, Bournemouth	9	Dennis Trident	East Lancs Cityzen
Yorkshire Terrier	15	Dennis Dart SLF	East Lancs Spryte
Zak's, Birmingham	2	Dennis Dart SLF	Plaxton Pointer 2

a – delivered 1998-1999
b – delivered 1999-2000

Forward orders

Orders announced during 1999 for delivery in 2000:

Arriva	20	Volvo B7L	Alexander ALX400
	27	Dennis Trident	Alexander ALX400
	34	DAF DB250	Alexander ALX400
	51	Dennis Dart SLF	Alexander ALX200
	65	Dennis Dart SLF	Plaxton Pointer 2
	12	DAF SB220	Alexander ALX300
	26	Volvo	Alexander ALX300
BC Transit, Canada	46	Dennis Dart SLF	Plaxton Pointer 2
Brighton & Hove	20	Dennis Trident	East Lancs Lolyne
Bullocks, Cheadle	4	Volvo B7L	Plaxton President
Bus Eireann	52	Volvo B10M	Plaxton Excalibur
Connex, London	27	Dennis Trident	Alexander ALX400
Dublin Bus	100	Volvo B7L	Alexander ALX400
	20	Volvo B6BLE	Wright
	2	Volvo B7L	Plaxton President
Dunn-Line	11	Volvo B10M	Plaxton
FirstGroup	75	Volvo B10BLE	
	55	Volvo B7L	Alexander ALX400
	60	Volvo B6BLE	Wright Crusader II
	20	Volvo B7L artic	Wright Fusion
Flights, Birmingham	24	Volvo B10M	Plaxton
Go-Ahead London	126	Volvo B7L	Plaxton President
	46	Volvo B7L	Alexander ALX400
	13	Dennis Trident	Plaxton President
Go-Ahead North East	27	Dennis Trident	East Lancs Lolyne
	37	Dennis Dart SLF	Plaxton Pointer 2
KMB	110	Dennis Trident 3	Duple Metsec
	60	Neoplan Centroliner dd	
London United	45	Volvo B7L	Alexander ALX400
	25	Volvo B7L	Plaxton President
Lothian	5	Volvo B7L	Plaxton President
	50	Dennis Trident	Plaxton President
MTL	55	Dennis Dart SLF	Marshall Capital
National Holidays	10	Volvo B10M	Plaxton Panther
New World First Bus	40	Volvo Olympian	Alexander ALX500
	30	Neoplan Centroliner dd	
Parks, Hamilton	20	Volvo B10M	
Parry, Cheslyn Hay	11	Neoplan Starliner	
Redwing, London	20	Setra 315	
Shearings, Wigan	30	Volvo B10M	
Stagecoach	170	MAN18.220	Alexander ALX300
	106	Dennis Trident	Alexander ALX400
	34	Dennis Dart SLF	Alexander ALX200
Translink (Ulster)	4	Mercedes O405N	Mercedes
	2	Mercedes O405GN	Mercedes
Wallace Arnold, Leeds	51	Volvo B10M	Plaxton
Wellglade group	38	Optare Excel 2	
	26	Optare Solo	
Yorkshire Coastliner	7	Volvo Olympian	Alexander Royale
Yorkshire Traction	12	Volvo B6BLE	East Lancs
Yorkshire Terrier	11	DAF SB220	East Lancs

Dennis and Plaxton supplied new buses to Canada - the first new British built buses to be bought by a Canadian fleet since Daimler Roadliners in the 1960s. These were air-conditioned Pointer/Darts for BC Transit. More will follow in 2000, along with Trident double-deckers.